Mechanics for Movement

D1825735

Mechanics for Movement

Notes for Physiotherapy Students

Françoise A Macdonald
Aberdeen Hospitals School of Physiotherapy

Bell and Hyman
London

Published by
BELL & HYMAN LIMITED
Denmark House
37–39 Queen Elizabeth Street
London SE1 2QB

First published in 1973 by
G. Bell & Sons Ltd.
Reprinted 1978

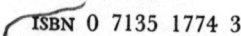
ISBN 0 7135 1774 3

PRINTED IN GREAT BRITAIN BY
REDWOOD BURN LIMITED, TROWBRIDGE AND ESHER

Contents

Acknowledgment

To Miss R. E. J. Lane and Miss R. A. Burnett, respectively Principal and Assistant Principal of the Aberdeen Hospitals School of Physiotherapy, Miss A. Dey, and my husband Donald, without whose encouragement, humour and positive help this book would never have materialized.

1   Introduction to Mechanics

MECHANICS is the science of matter and force, and therefore deals with forces acting on bodies, and with the result of these forces in terms of movement and equilibrium.

The application of mechanics to the living human body is referred to as *Biomechanics*.

Mechanics is divided into:

1   *Statics;* concerned with bodies in balance.
2   *Dynamics;* concerned with bodies in motion.

MATTER is anything which occupies space and has weight. The three states are solid, liquid and gas.

Matter is made up of molecules, which are the smallest particles into which it can be divided without destroying its characteristic properties. Molecules are made up of atoms which in turn are made up of protons, neutrons, and electrons.

All matter can change into all three states. The difference in these states is the amount of movement the atoms and molecules make, and this depends on the temperature.

In a SOLID: The molecules are closely packed together. They do not move in relation to each other, but vibrate together while remaining in their fixed pattern, therefore, it is very difficult to compress solids into a smaller volume. (Figure 1)

In a LIQUID: The molecules are about as closely packed as in the solid state, but they slip against each other in all directions. The hotter the liquid the faster the molecules move. Because the molecules slide so easily, a liquid has no shape and therefore takes the shape of its container. However, because its molecules are tightly packed, it cannot be pressed into a smaller volume, except under great pressure. (Figure 2)

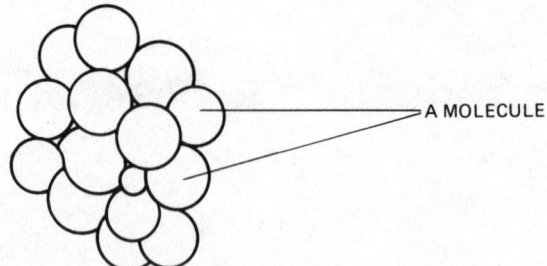

FIG. 1. *Molecules in a solid*

FIG. 2. *Molecules in a liquid*

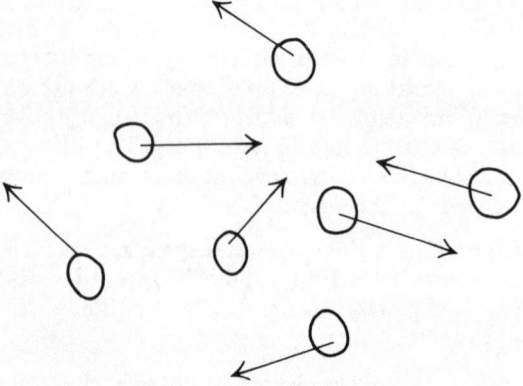

FIG. 3. *Molecules in a gas*

In a GAS: The molecules are moving so fast that they have broken away from each other. They take up a great deal more room than when solid or liquid, and there is an enormous amount of space between them. It is easy to compress a gas (it is just a matter of reducing the space between the molecules). Like a liquid, a gas has no shape, but unlike a liquid it will spread out to fill any container completely. (Figure 3)

Changes in matter are of three kinds:

1 *Physical* in which the composition of the molecules of the substance is not changed. (For example, water into ice.)

2 *Chemical* in which the composition of the molecules is changed and new substances with new properties are produced. (For example, the rusting of iron which produces iron oxide.)

3 *Nuclear* in which new materials are formed by changes in the identity of the atoms themselves.

ENERGY is the ability to do work.

> Work = weight × distance moved.
> (Measured in lb.wt. × ft., for example.)

The speed at which the work is done does not affect the total work. The work done is the same when a given weight is moved a certain distance, irrespective of whether the movement is slow or quick.

However, the POWER is different as the rate of doing work is different. POWER = work done in 1 sec. One good horse was found to do 550 ft.lb.wt. of work in one second, and this amount of work was therefore called one horse power (1 h.p.).

Energy comes in various forms:

Mechanical
Chemical
Light
Magnetic
Atomic

Any one form can be converted into another with a certain amount of waste. Energy can never be destroyed or created. This is known as the *Law of Conservation of Energy*. Therefore, the total amount of energy in the whole universe has always been the same and always will be the same.

The Relation between Matter and Energy

Matter can be changed into energy. For example, an atomic explosion results when a minute amount of matter is converted into a vast amount of energy. The energy has not been created out of nothing, but from matter. Therefore, matter is just another form of energy.

Mechanical energy is either Kinetic (associated with moving objects), or Potential (stored up) energy.

FORCE is a push or a pull which does not necessarily produce motion.

Therefore, it is that which produces or prevents motion or has a tendency to do so.

Force can therefore produce *movement* as when a ball is pushed, or it can produce *strain* as when sitting on a chair which might creak, or it can produce *distortion* as when pressing on putty.

WEIGHT is the *force* exerted by *matter*.

Weight is the force with which matter is attracted towards the earth.

When one is finding the weight of an object one is measuring the attraction of the earth for that object.

The weight of an object depends on two things:

1   The mass or quantity of matter it contains.
2   The amount of gravitational attraction the earth has for it.
   $W = M \times g$ where W = weight   M = mass   g = force of
                                                         gravity.

MASS is the measure of the quantity of matter which a body contains. It is the number of molecules and atoms it contains and therefore remains constant irrespective of any change in volume or geographical situation. Mass is measured in pounds/grams. These are absolute units, because they are not dependent on the force of gravity.

GRAVITATIONAL ATTRACTION will change according to where the object is. It will weigh less at the top of a mountain (and even less on the moon!) than in a valley.

To measure force one has to know both its magnitude (or size) and direction. It is therefore known as a *Vector Quantity*. A Vector Quantity is one which has magnitude and direction, whilst quantities which have no particular direction but only magnitude, e.g. volume or density, are called *Scalar Quantities*.

Vector Quantities can be represented by an arrow which is called a Vector (Figure 4). The head of the arrow indicates the direction

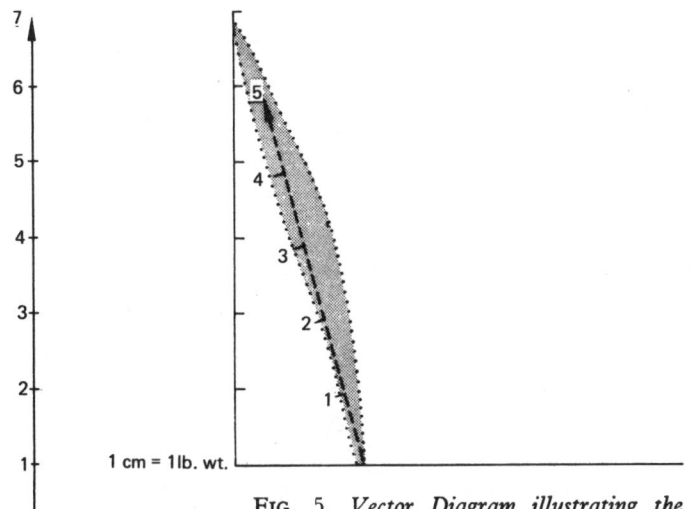

FIG. 5. *Vector Diagram illustrating the pull of the biceps muscle on the radius*

FIG. 4.   *A Vector*

(i.e. the pull). The length represents the magnitude of the force (in this case, 7 units of force). The end of the arrow represents the point of application of the force.

Any convenient unit may be used to represent one pound of force, as long as the same unit is used when several forces are represented on one diagram. Figure 5 represents the pull of biceps muscle on the radius. The vector force drawn represents a pull of 5lb. in an upward direction.

Types of Forces

    1   Linear forces.
    2   Parallel forces.
    3   Concurrent forces.

Linear Forces are those which all occur along the same action line as in a tug of war.

1   A single linear force acting on a body which is free to move causes movement in the direction of the force. (E.g. pushing a marble.)

2   Two equal forces acting at a common point and in opposite

directions will result in a state of equilibrium (Figure 6). The book and table are in equilibrium because the downward force of the book is opposed by an equal but upward force from the table (Newton's 3rd Law). Both forces have the same action line.

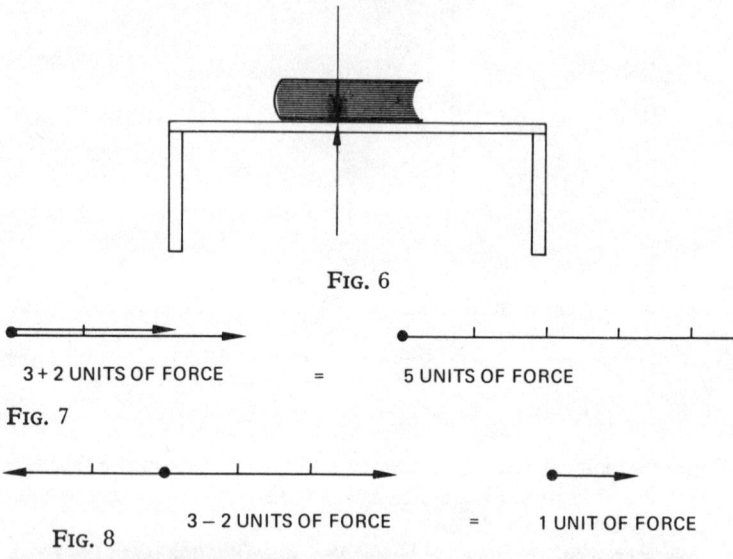

FIG. 6

3 + 2 UNITS OF FORCE   =   5 UNITS OF FORCE

FIG. 7

FIG. 8   3 − 2 UNITS OF FORCE   =   1 UNIT OF FORCE

3 Two forces acting in the same direction and at a common point are equal to a single force acting in that direction whose magnitude is equal to the sum of the magnitude of the individual forces. (Figure 7)

4 Two unequal forces acting at a common point and in opposite directions will result in movement in the direction of the greater force. The magnitude of this force will be the difference between the greater and smaller forces in opposition to each other. (Figure 8)

Parallel Forces are those which lie parallel, and in the same plane to each other but do not act along the same action line. In Figure 9, children on a see-saw exert downward forces which are parallel to each other. Their combined weights must be opposed by the upward force at the axis of the board. The resultant of the two parallel forces in the same direction equals the sum of the forces $= B + C$. In equilibrium, $B + C = A$. Anatomically when the forearm is flexed

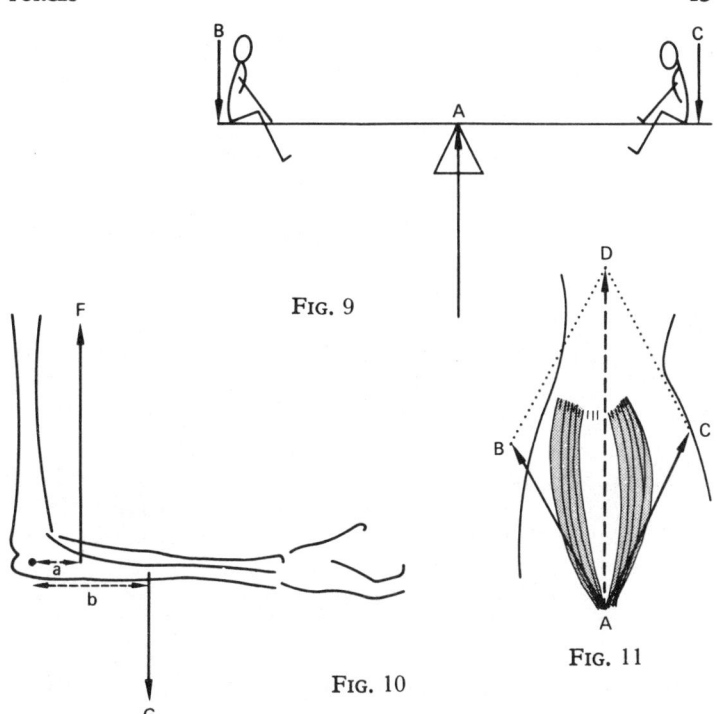

Fig. 9

Fig. 11

Fig. 10

to horizontal, gravity pulls downwards and the biceps muscle pulls upwards. These two action lines are parallel and opposite. (Figure 10)

When two equal parallel forces act on a body in opposite directions they tend to make it rotate. This is described as a *Couple* and can be seen acting when both hands are on the steering wheel of a car. In the human body, the upper fibres of trapezius and the lower fibres of serratus anterior, rotate the scapula upwards by acting co-operatively as a couple with the lower fibres of trapezius.

Concurrent Forces are those which meet at a point. This is very common in the human body where two muscles will act in different directions from a common point to produce one movement. In Figure 11 the anterior and posterior fibres of deltoid will flex and extend the arm in a sagittal plane when working singly. However, working together, pulling on the deltoid tubercle of the humerus, they will abduct the arm in the frontal plane. The resultant force is calculated from the parallelogram law. (See pages 15 and 16.)

Examples of Concurrent Forces Acting on the Body

1   Clavicular and sternal heads of pectoralis major muscle act together horizontally to adduct the arm.

2   Anterior and posterior fibres of deltoid muscle.

3   Two heads of gastrocnemius pulling in lateral and medial directions exert an upward force on the Achilles' tendon.

4   Anterior and posterior parts of gluteus medius in hip abduction.

5   Evertors and invertors of the foot in plantar and dorsiflexion.

6   Radial and ulnar deviators of the wrist in wrist flexion and extension.

Obviously, it can be seen now that two or more forces can act simultaneously on the same object. Then it may be useful to find the magnitude and direction of a single force which could produce the same effect as all the forces acting together. This is known as the *Resultant Force* and this force must produce exactly the same effect as all the forces for which it is substituted.

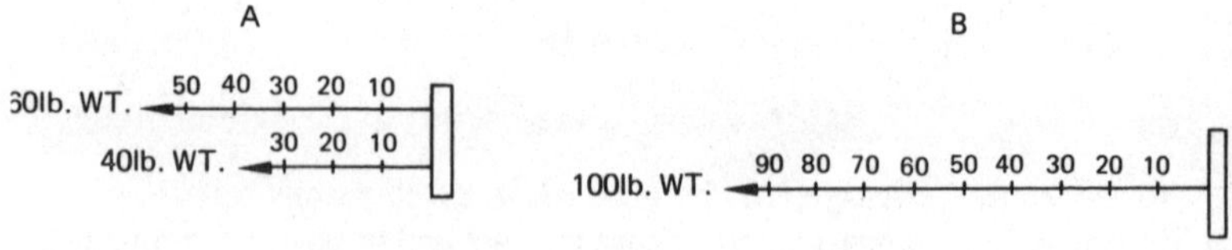

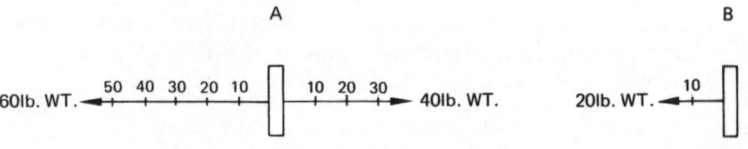

FIG. 12

FIG. 13.   *B is the resultant of A*

The Resultant of Forces Acting in a Straight Line (Figure 12)

If one boy pulls on a rope with a force of 40 lb. and another boy joins him and pulls with a force of 60lb. in the same direction, the resultant force equals the sum of the two separate forces, i.e. one man pulling with a force of 100lb. would have the same effect.

If one boy pulls in one direction with a force of 40lb. and the other boy pulls on the same point in the opposite direction with a force of 60lb. then the resultant force will be the difference between the two forces, i.e. 20lb pulling in the direction of the greater force will have the same effect. (Figure 13)

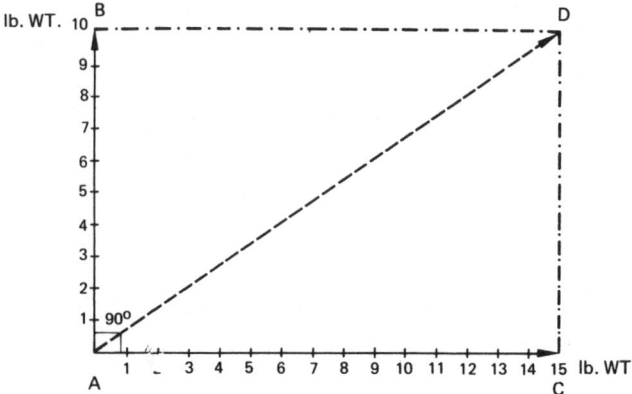

Fig. 14

The Resultant of Forces Acting at Right Angles (Figure 14)

Referring to Figure 14, two vectors have been drawn to represent two forces acting at right angles on one object at A. AB = force of 10lb. AC = force of 15lb. Force AB will tend to pull to B. Force AC will tend to pull to C. When both forces are pulling simultaneously the object will move along the diagonal of the parallelogram of which the two forces are sides.

This produces a rectangle divided into two right-angled triangles. According to Pythagoras, in a right-angled triangle the square on the hypotenuse is equal to the sum of the squares on the other two sides.

Therefore $AD^2 = AC^2 + CD^2$
Therefore $AD^2 = 10^2 + 15^2$
Therefore $AD = \sqrt{325} = 19.03$lb. Resultant Force.

Therefore the resultant of two forces acting upon a given point at an angle is equal to the diagonal of a parallelogram of which the two force vectors are sides. This is known as the *Parallelogram of Forces.*

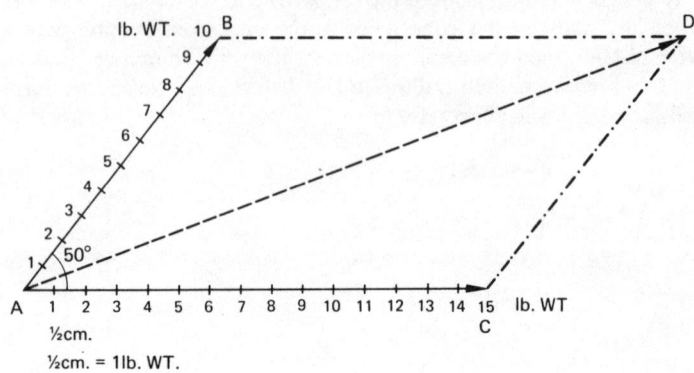

FIG. 15

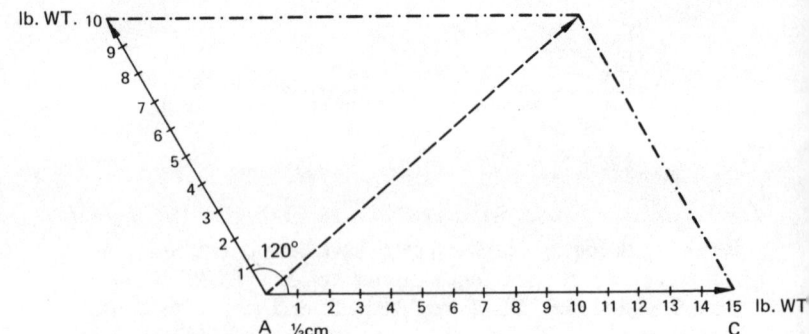

FIG. 16

The Resultant of Forces at Any Angle

The angle between two forces acting at a point is generally not a right angle. In Figure 15 two vectors have been drawn to represent two forces acting at 50° to an object, A. The other two sides of the parallelogram are then drawn. The resultant will be AD. This will be calculated either by trigonometry or by measuring the length of the resultant and converting it to lb.weight. AD = 11½cm. =

approx. 23lb.weight. Compare this with the same two forces acting at a point at an angle greater than 90°. (Figure 16). In the diagram, the forces are acting at 140° to each other. AD = 5cm. = approx. 10lb.weight.

When three or more forces act on the same point their resultant is found by first finding the resultant of two of the forces, then use this diagonal plus the third vector to get the resultant of the three concurrent forces.

When forces act on a point, the *Equilibrant Force* will be equal to the Resultant Force but in exactly the opposite direction so producing equilibrium or preventing movement.

The opposite of composing two or more component forces into one force, the resultant, will be to separate one force of given magnitude and direction into two components. This is known as the *Resolution of Forces*.

Resolution of Forces

The resolution of forces is the separation of a single force into two forces acting in definite directions upon the same point. Usually these two forces act at 90° to each other.

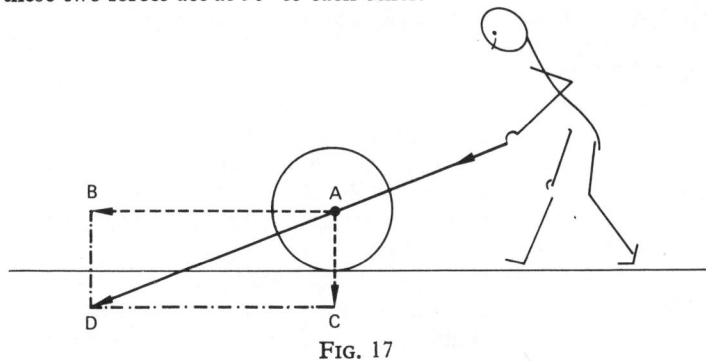

FIG. 17

In Figure 17 where a roller is being pushed forwards, the force AD (that force applied to the handle and transmitted to the axle) is resolved into two components:

AB, which pushes the roller forwards.
AC, which acts perpendicularly and tends to push the roller into the ground.

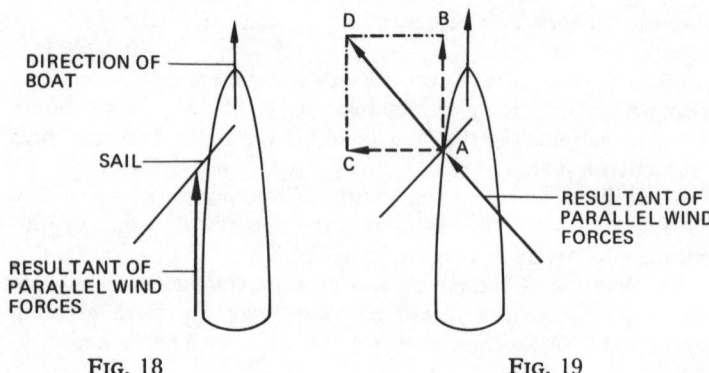

FIG. 18   FIG. 19

In Figure 18 representing a sail-boat moving with the wind, the full force of the wind drives it forwards. However, when the boat moves at an angle to the wind as in Figure 19, AB pushes the boat forwards, and AC tends to tip the boat sideways. This will be applied to the force of a muscle acting on a bony lever.

Centre of Gravity

Consider an object of any shape. Every part of it has weight and as such will be attracted downwards by earth, and every part can be represented by a vector force. All these forces will be parallel to each other. The total weight of the object will be the resultant of all the forces. The point of application of the resultant will be the Centre of Gravity of that object.

The Centre of Gravity of an object is that point at which all the weight of the body may be considered to be concentrated, and about which all the parts exactly balance. In Figure 20 the equilibrant will be the force needed, applied immediately above the Centre of Gravity, to lift the stone without rotation. This will be equal and opposite to the Resultant. If a force is applied at A to turn the object over, it will pivot at point B. The force at A acts on the lever arm AB to turn the object over in a clockwise direction. The weight of the object acting at C acts in an anti-clockwise direction with lever arm CB. The force at A is applied to a longer lever arm AB than that of the object CB, therefore, less force will be needed to turn the

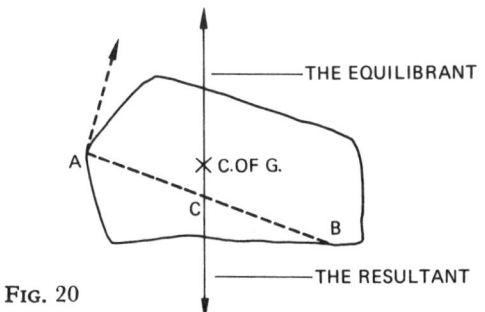

FIG. 20

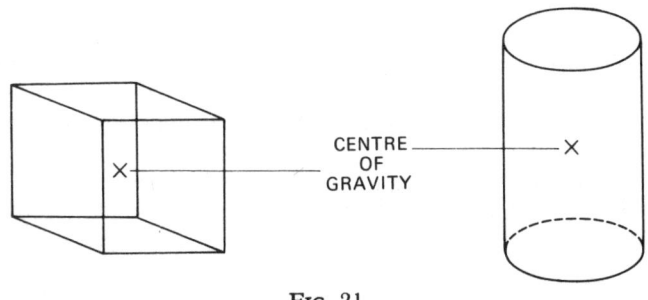

FIG. 21

object than the weight of the object itself, i.e. less force is needed to turn the object over than to lift it. This can be related to turning a patient over on his side by rolling him rather than by lifting him.

In a block or cylinder in which the mass is symmetrically distributed the centre of gravity is at the geometric centre of the object. (Figure 21). However, if the distribution of mass is asymmetrical as in the human body the centre of gravity will be nearer the larger and heavier end.

In man the centre of gravity is within the pelvis in front of the upper part of the sacrum when standing in the anatomical position, but this will vary according to build, sex and age. It will also vary within any given person when the arrangement of the segments of the body shift as in running, sitting, etc. Since this point represents the centre of the total mass, it will shift when weight is added to, or subtracted from, some part of the body, as with the addition of a brace or cast, or following an amputation of an extremity. The important thing to remember is that a rigid object behaves as if its entire mass were acting, or being acted upon, at its centre of gravity.

The Line of Gravity

This is an imaginary vertical line passing through the centre of gravity. In the human body it passes through the body of the second sacral vertebra down to a point between the feet, when standing in the anatomical position.

The Base of Support

This is the area through which the line of gravity of an object must fall, in order to maintain equilibrium.

Stability

A book may be pushed far out over the edge of a table but it will not topple until its centre of gravity is no longer supported. In other words, the object behaves as if its gravitational pull concentrates its entire force at the single point of its centre of gravity. Thus the book falls as soon as its centre of gravity passes over the table edge despite the fact that a large mass of the book is still supported. (Figure 22)

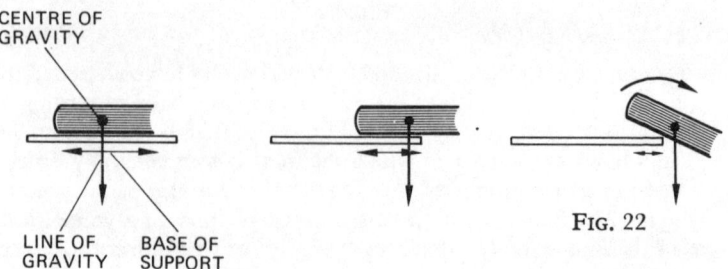

CENTRE OF GRAVITY

LINE OF GRAVITY   BASE OF SUPPORT

FIG. 22

The base of support is therefore very important to the stability of an object and for an object to remain upright the vertical line from the centre of gravity, that is its line of gravity, must fall inside its base of support. From this it can be seen that the larger the base, the easier it will be to maintain the centre of gravity over it. The smaller the base the more difficult.

In a man whose weight is entirely supported by his feet the base of support includes not only his two feet but the space in between. Thus in the upright human body it is least stable when the feet are parallel and close together. When the feet are separated, the base is widened and stability improves. (Figure 23). However, if the feet

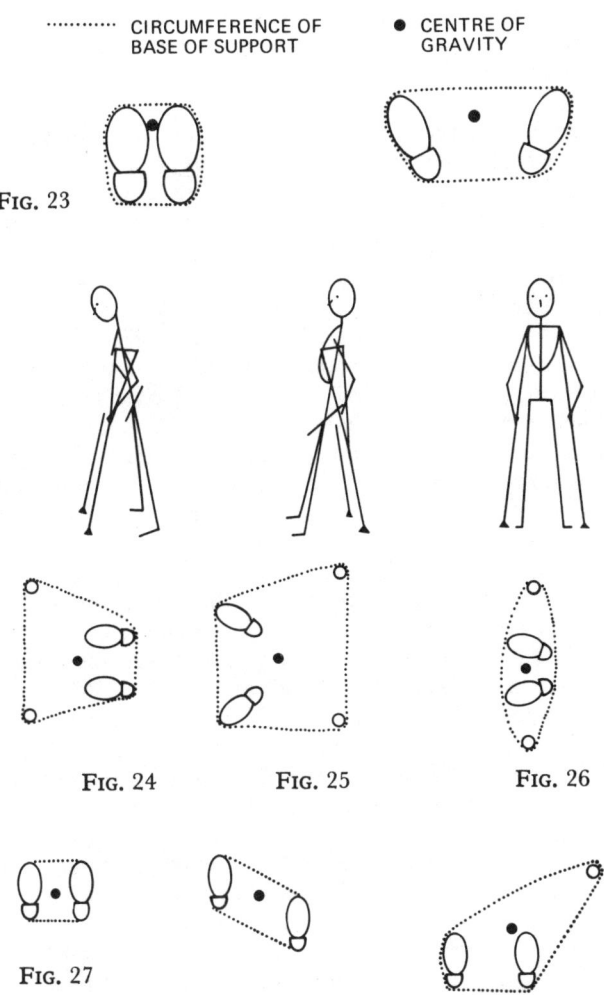

·········· CIRCUMFERENCE OF   ● CENTRE OF
             BASE OF SUPPORT   GRAVITY

FIG. 23

FIG. 24   FIG. 25   FIG. 26

FIG. 27

are farther apart than the width of the pelvis the legs will be in a slanting position. If this is accompanied by insufficient friction between the feet and the supporting surface (e.g. a patient standing on a slippy floor wearing leather-soled shoes) then this does not make for greater stability.

The stability of the physiotherapist is very imporant, and she will try to maintain her centre of gravity over a wide base of support whenever possible. When giving bodily support to a patient the centre of gravity will now be at the combined weight centre of the two people.

A man on crutches will further increase his base of support and thus increase his stability. He can do this in different ways. In Figures 24 and 25 his antero-posterior position is greatly improved. In Figure 26 his lateral stability is improved but it must be remembered that as soon as the crutches are lifted off the ground the patient's stability, if anything, is decreased as his centre of gravity has been raised by the addition of the weight of the two crutches. The use of a cane will also increase stability. (Figure 27)

In the spine, the normal curves in the cervical vertebrae are convex forwards, in the thoracic vertebrae are concave forwards, in the lumbar vertebrae are convex forwards and the sacral and coccygeal vertebrae are concave forwards. The line of gravity will fall so that the length and angular value of the convex curves will be equal to that of the concave curves. If for any reason any of these curves are altered there will be compensation in the other curves to maintain equilibrium.

Balancing the centre of gravity over the feet becomes a critical problem in patients with spinal cord lesions and those with muscle weakness of the trunk and lower limbs. A paraplegic must be able to balance the movements of the head and shoulders which he can control, through his muscles, and his pelvis which he cannot control. He must be able to master the interplay in position between the pelvis and his head and shoulders. He must be able to control his head by moving it backwards or forwards to help place the pelvis in the right position so as to maintain the centre of gravity within his base of support, and place the lower limb mass correctly as for walking, climbing stairs, etc. The physiotherapist must help him to work out the most stable position at rest and in movement. In Muscular Dystrophy the absence of strength in the trunk muscles often causes the patient to arch his spine in an exaggerated lordosis to balance in the upright position. Any touch might shift his centre of gravity outside his base of support and therefore he must be allowed to walk slowly and delicately.

The speed of movement is closely associated to the requirements of balance. It is far easier to balance on a quickly moving bicycle than when moving slowly. Therefore, patients with a precarious sense of balance hurry along in order to decrease the requirements of lateral stability.

2 Equilibrium

A RIGID body is in equilibrium if the forces acting on it do not tend to move it in any direction or to rotate it about any axis. The resultant of all the forces acting on the object will be zero. A body can only be in equilibrium under the action of two forces if they are equal and work in opposite directions in the same action line. If a body is in equilibrium under the action of three forces the resultant of any two of these forces must be equal and opposite to the third force. This can only happen when the three forces act in a single plane and then their lines of action are either parallel or meet in a single point.

However, balanced objects possess varying degrees of firmness or steadiness in their positions. There are three types of equilibrium:

1 Stable.
2 Unstable.
3 Neutral.

Stable Equilibrium

An object will be in stable equilibrium if, when it is displaced from its original position at rest, it will be restored to that position by the forces acting on it. An object in stable equilibrium cannot be overturned without first raising its centre of gravity. In Figure 28 there are two bricks. Both bricks have the same volume and dimensions and both are in stable equilibrium but their degree of stability differs.

Brick A

To overturn this brick, its centre of gravity must be raised to point E, for only when it passes this point will the line of gravity fall outside the original base of support. The brick must now find a new base of support, and will come to rest in the upright position lowering its centre of gravity from E to B.

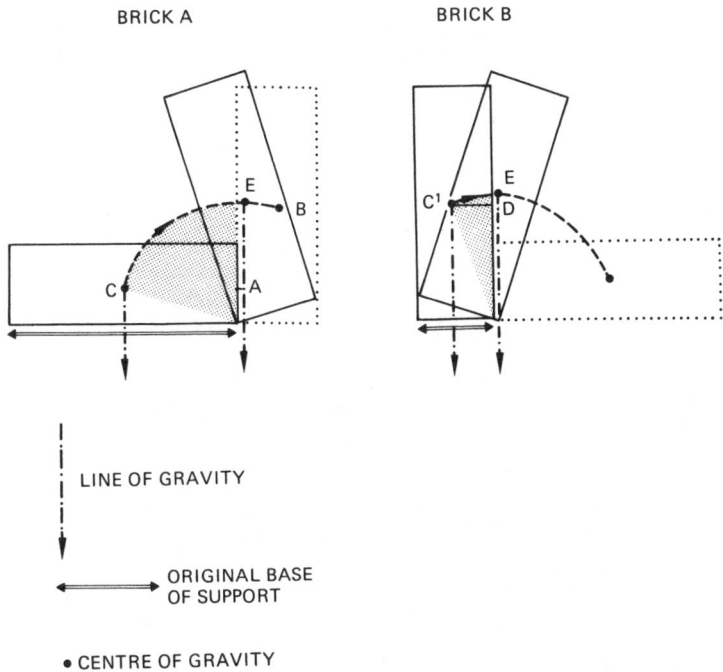

BRICK A   BRICK B

LINE OF GRAVITY

ORIGINAL BASE
OF SUPPORT

• CENTRE OF GRAVITY

FIG. 28

Brick B

This is the same brick as A but is standing on its end. To overturn this brick its centre of gravity must be raised from C^1 to E. The centre of gravity will no longer fall over the original base, and to find a new base the brick falls on its side, thus lowering its centre of gravity, and comes to rest on its side.

From this it can be seen that the centre of gravity of brick A must be raised the distance from A to E before it must change its base of support and be turned on its end. The centre of gravity of brick B must be raised from D to E before it falls on its side than upended. The angle through which the centre of gravity must tilt to lose balance is much greater in A than in B, and the work done to achieve this is also greater as the effort needed is also greater. A heavier brick would be more stable in both positions as it would take more effort and therefore more work to topple it.

The stability of an object will depend on:

1 The size of the base of support.
2 The height of the centre of gravity.
3 The horizontal distance between the centre of gravity and the
   pivoting edge.
4 The weight of the object.

Unstable Equilibrium

An object will be in unstable equilibrium if when displaced from
its original position at rest by forces, it will be further displaced by
these forces. For example, a tightrope walker or an egg on its end.
(Figure 29). As soon as the slightest movement occurs the centre of

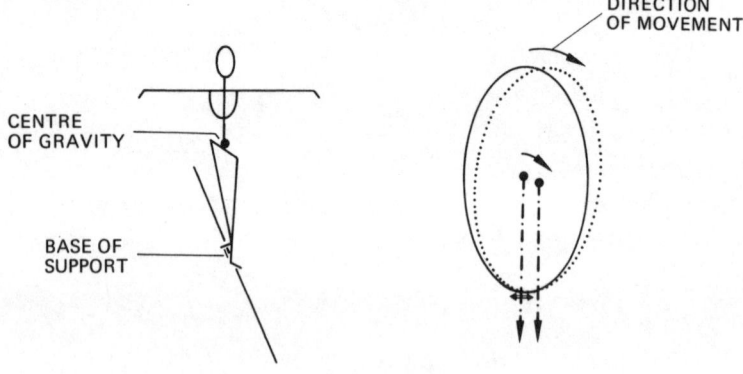

FIG. 29

gravity falls outside the base of support. The centre of gravity is
immediately lowered and the object falls. Thus the narrower the
base and the higher the centre of gravity the more unstable a body
will be.

However, a solid possessing a small base can be stood up on its
end with difficulty. A pencil may be stood on its end, a pencil stub
more easily, a long stick with great difficulty and only for a very
short time. Thus the lower the centre of gravity the more stable the
object is. (Figure 30). On the other hand the long stick can be stood
on end on its point almost indefinitely if the finger supporting it is
kept moving, so as always to bring the point of support beneath the

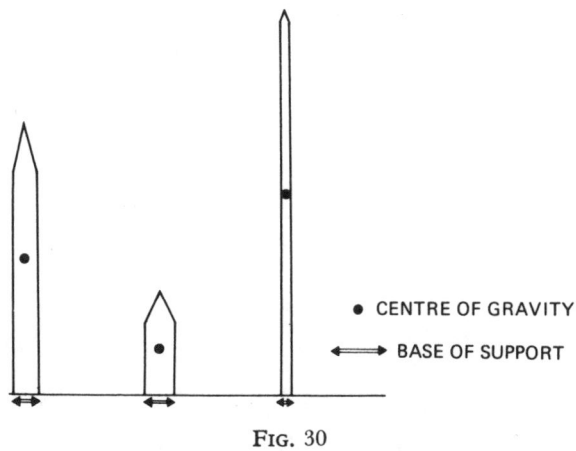

FIG. 30

centre of gravity. A long pencil can also be balanced on its point but with greater difficulty, and a pencil stub is practically impossible to balance on its point. When the long stick is being balanced, because its centre of gravity is high, it moves much more slowly from its position of equilibrium than it does when the centre of gravity is low as in the pencil stub. If the top of the stick is weighted with lead it will be even easier to balance than the stick alone. Thus in cases of dynamic balance, the higher the centre of gravity within limits, the easier is the ability to balance. This illustrates another possible method of balance by means of the process of perpetual adjustment. The human body, except in lying, fully supported, maintains balance by this process of perpetual adjustment.

Neutral Equilibrium

An object is in neutral equilibrium if when displaced from its original position of rest by a force, it will come to rest in any position as its centre of gravity is neither raised nor lowered when it is overturned. For example, a ball or cylinder or cone lying on its side. (Figure 31)

The human body is in a state of neutral or dynamic equilibrium when moving. Balance is maintained by a continual adjustment of the base of support. In the adult this is performed as a reflex action but if one looks at the staggering walk of the one-year-old toddler trying to maintain balance it will be seen that this technique has to

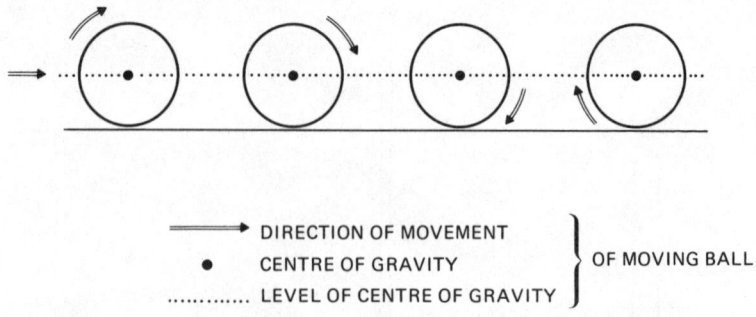

DIRECTION OF MOVEMENT
• CENTRE OF GRAVITY   OF MOVING BALL
.... LEVEL OF CENTRE OF GRAVITY

Fig. 31

be learned. If the centre of gravity of the body moves more rapidly than the feet, thus allowing the line of gravity to fall outside the base of support, then a new base must be found or the body will fall over. This might happen when jumping off a moving bus. The bus has imparted kinetic energy or momentum to the body, and the body when it first makes contact with the ground, is thus moving forward with the speed of the vehicle. It will continue to move forward at this speed as there are negligible forces available to stop it, apart from those exerted by the feet. If this initial speed is greater than the maximum speed at which that person can run, his centre of gravity will overtake his body and he will fall forwards.

The human body can never be in a state of stable equilibrium (as stable equilibrium implies that no movement is necessary to maintain the position), except when, fully supported in lying, no interplay of muscle tone is necessary to maintain the position. Otherwise there is always an amount of swaying to maintain any position which is caused by interplay of muscle tone which is entirely regulated by sensory feedback. The human body is therefore always in a state of dynamic equilibrium. There will always be interplay of muscle tone to keep the centre of gravity over the base in any position the body might adopt, whether it is obviously moving or remaining still. Equilibrium is maintained by a negative feedback where the body, because of sensory stimuli, is aware of being in a position of imbalance, and will correct this. The body will then overbalance in the opposite direction and be corrected again, and so on, thus producing a sway. If the body were to maintain equilibrium by a positive feedback it would be rigid and there would be no sway. Where there is loss of proprioceptive sensory feedback, or of motor power, balance will be impaired. It may be possible for the patient to substitute for the proprioceptive loss by the use of other sensory mechanisms, particularly the eyes.

An object only retains its equilibrium as long as its centre of gravity falls within its base of support. Therefore the nearer the line of gravity falls to the centre of the base of support the greater the stability. The farther away the line of gravity the less the stability. Once the line falls outside the margin of the base of support a new base must be found. Normally this is done subconsciously but in a patient who has lost this ability through illness or a disease that alters his centre of gravity, as in the hemiplegic or amputee, then this must be re-educated.

Principles Relating to the Maintenance of Equilibrium

1 Maintain an adequate base of support.
2 Lower the centre of gravity when possible.
3 Keep the centre of gravity well centred over the base of support.
4 Increase the size of the base of support in the direction of the force or movement.

When a body is at rest the resultant of all the forces acting on that body must be zero to maintain equilibrium. With reference to Figure 28 the brick is in equilibrium in both positions because the force of gravity pulling the brick down is equal and opposite to that of the ground pushing it up.

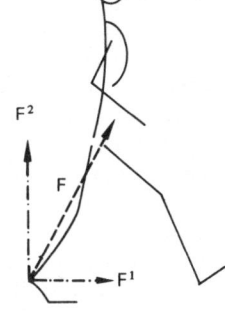

When a body is in motion, equilibrium is maintained when clockwise and anti-clockwise rotations cancel out. In Figure 32 the force F is used to push off with his back foot. This force which passes through his centre of gravity, is resolved into two components:

F^1 which tends to rotate him backwards.

F^2 which tends to rotate him forwards.

Fig. 32

To be in dynamic equilibrium when walking, the force tending to rotate him backwards must be exactly counteracted by the force tending to rotate him forwards, and this must happen at all points in the walking cycle.

Friction

In considering the stability of a human body relative to its base of support when standing, it was agreed that stability could decrease once the feet were placed farther apart than the width of the pelvis if there was not sufficient friction between the feet and the floor to give adequate hold. Friction is the force which tries to stop one object from sliding over another. How does it do this? When a weight rests on a table and the hand pushes against it slightly, it does not move. It is still in equilibrium. The force of the hand is being counteracted by an equal and opposite force on the weight. This opposing force can only be exerted by the table and must be directed horizontally. This force is the Force of Friction, which prevents the weight from being accelerated by the force of the hand.

Now let the hand push a bit harder; still the weight remains at rest. The force of friction has the remarkable quality of adjusting its magnitude exactly to that of the force tending to produce motion. However, this is not true indefinitely. If the hand pushes with sufficient strength the weight will move, but with an acceleration corresponding to a force smaller than that applied. The net force acting to produce movement will be the difference between the maximum force of friction and the force applied. The friction exhibited which tends to prevent movement is known as *Static Friction*.

A moving body is also retarded by friction. Experiments show that the retarding force on sliding objects is in general slightly smaller than the maximum force of friction on stationary ones. This is known as *Kinetic Friction*.

The rolling of a wheel is also impeded by a force but for different reasons. No surface is completely rigid and a wheel will make an impression on it. As the wheel moves forwards it has to overcome a slight hump in front of it and the effect is the same as if it were rolling uphill. This is known as *Rolling Friction*.

There are various suggestions as to how the force of friction works:

1   Irregularities in the surface of the objects being rubbed together tend to interlock and therefore offer resistance to movement. The more polished the surfaces are the less the friction between them, but this only applies to a certain degree.

2   Electrical forces between the surfaces holding them together.

3   Adhesions of the molecules of one surface to the surface of the other. The molecules that are rubbed off as one surface slides over the other give a varying amount of resistance.

Friction is usually necessary for movement to start. To walk, there must be adequate friction between the soles of the feet and the floor so that the feet will not slip. The greater the friction between the supporting surface and the parts of the body in contact with it, the greater the body stability will be. If the supporting surface offers little friction as in the case of icy pavements, slippery floors, fallen wet leaves, etc., then footgear can compensate for it in the form of rough-soled shoes, hob-nailed boots, studded golf, rugger or football boots to name but a few. If friction is minimal the body will increase its stability by increasing its base of support. This is of even greater importance when the body is in motion or is subjected to an external force. In movement, a heavy human body stands a better chance of keeping its footing than a lighter one when an external force is applied, because the force of friction is directly proportional to the force (i.e. weight) pressing the two surfaces together. However, it is obviously not practical to change one's weight according to conditions but there are two ways in which to withstand this factor of external force:

1   Inclining the body in the direction of the oncoming force.

2   Widening the base of support in the direction of the oncoming force.

For example, a person will incline his body into a strong wind, and he will widen his base of support to resist movement when a bus starts. A skater has no heel strike or normal push off as there is insufficient friction for these to be effective. He will 'push off' with the edge of the blade thrusting at right angles to its long axis. He will maintain his equilibrium, by inclining his body in the direction of movement, and by widening his base of support.

Examples of the Practical Use of Friction

1 Friction is essential for normal push off in walking.
2 Friction is necessary for a wheelchair to be propelled along the floor. (It would be very dangerous if it were allowed to slide.)
3 Friction is produced between the tips of crutches and sticks and the floor.
4 Friction is used to provide resistance in graded exercises using the stationary bicycle.
5 Friction is made use of in brakes, and locks on castors.
6 Friction is necessary between the patient and the bed to apply lumbar traction.

Examples where Friction is Minimised

1 In suspension when the body or part of the body is suspended a sliding movement only occurs at the point of contact of the rings and hooks of the suspensory unit as movement is performed. Both these surfaces are rounded and therefore the point of contact is small so friction is virtually eliminated.

2 In massage, friction is minimised by the use of lubricant such as talcum powder or oil.

3 Friction is minimised by lubricating moving joints in any apparatus.

4 Friction is minimised when giving passive movements, by applying traction to the joints concerned.

Newton's Laws on Motion

These are three laws which relate force and motion.

Motion or movement is defined as a continuous change of place or position. There are two kinds of movement:

1 Translatory or linear movement.
2 Rotatory or circular or angular movement.

Linear or Translatory movement is movement of an object in a straight line, with all its parts moving in the same direction, over the same distance, at the same speed. E.g. a stone sliding on ice, or an object on a conveyor belt.

Rotatory or Circular or Angular movement takes place about a fixed or relatively fixed axis, and is therefore dependent on a system of levers. For example, flexion of the shoulder joint is a rotatory movement where the joint is the axis of movement and the arm is the lever. In this situation every particle of the arm will not move in the same direction simultaneously, nor will every part cover the same distance or travel at the same speed.

A body can perform both linear and rotatory movements simultaneously. The human body in walking performs rotatory movements at all joints involved to take the body as a whole in a particular direction, i.e. linear movement. Strictly speaking, this will not be true linear movement as each body segment will move with a different velocity. (Velocity by definition is the rate of movement in a particular direction. It is therefore a vector quantity.) The product of the mass of a body and the velocity with which it is moving is its *Momentum*, i.e.

$$\text{Momentum} = \text{Mass} \times \text{Velocity}$$

Force, as has been stated before, is that which produces or prevents movement or has a tendency to do so.

Newton's First Law of Motion (The Law of Inertia)

Newton's first law states that bodies at rest tend to remain at rest and bodies in motion tend to remain in motion. A ball rolling along level ground would continue to roll for ever in a straight line were it not for ground and air friction. Therefore, when a body is at rest or moving at a constant speed in a straight line, its velocity is constant. The velocity of an object does not change unless an unbalanced force acts on it. The velocity is changed when it is speeded up, slowed down, or its direction of movement is altered. When the velocity of an object is changed it will either be accelerated or decelerated. (Acceleration, by definition, is the rate of change in velocity and is measured in miles per hour per second, or, feet per second per second). Therefore an unbalanced force accelerates or decelerates (negative acceleration) an object. It is necessary to exert a force on an object to accelerate it because of the property of inertia which the body has. If a body is at rest its inertia will tend to keep it at rest. If a body is in motion its inertia will tend to keep it in motion. Inertia is concerned with a body's resistance to change in movement. It is proportional to mass and is, in fact, interchangeable with it. The mass of an object is the measure of its inertia, that is, its resistance to change in motion. To get an object moving, one must overcome both its inertia and ground and air friction and therefore more force is needed for this than to maintain a constant speed where only the frictional forces oppose the movement. However, to increase the speed, inertia has to be overcome once again.

Once the inertia of a body is overcome and movement is initiated, it is more economical to continue moving to avoid additional expenditure of force which would be required to overcome the inertia when stopping, starting or altering speed. Weak muscles may be unable to exert sufficient force to overcome inertia and therefore be unable to initiate a movement, yet may be able to produce movement or control movement if given assistance at the right time as in the use of *Sling Suspension*.

Newton's Second Law of Motion (Law of Acceleration)

This law relates force, mass and acceleration and states that an unbalanced force gives a body an acceleration in the direction of the force. This acceleration is proportional to the force and inversely proportional to the mass of the body. This will mean that a steady force will produce a uniform acceleration on the body it is acting on. If the force is doubled the acceleration will be doubled. If the mass is doubled then the force must be doubled to produce the same acceleration.

Newton's Third Law of Motion (Law of Reaction)

Newton's third law states that action and reaction are equal in magnitude but opposite in direction. If body A exerts a force on body B, then body B exerts an equal and opposing force on body A. For example, a book on a table exerts the same force down as the table exerts up, and when a gun is fired the momentum with which the bullet leaves the gun is equal and opposite to the recoil of the gun.

If a force acts upon a body and moves it, the force has done work upon the body. Work, therefore, refers to the overcoming of the resistance, and by doing so, moving the body through a certain distance. No mechanical work is achieved if in attempting to lift a heavy object, great energy is used, but no movement of the object occurs.

Work = force × distance moved, parallel to the force. E.g. if 1lb. is lifted 1ft. then 1ft.lb. of work has been done.

Energy is defined as the capacity to do work, and therefore the energy possessed by a body, is the measure of the work it is capable of doing. Mechanically there are two types of energy:

1 Kinetic energy.
2 Potential energy.

Kinetic energy is energy of motion. A body has kinetic energy because of its velocity. When starting a bicycle from rest to a specified speed, the muscles have exerted a force on the road, via the back wheel, which causes the bicycle to accelerate. That is, a force has been exerted over a distance, and work has therefore been done. Apart from the work done to overcome frictional and air resistance, all the work done will be stored inside the operator and the bicycle and is called kinetic energy. (I.e. energy of motion.) If pedalling is now stopped, the kinetic energy is gradually used up by doing work against friction and wind resistance, and the bicycle will gradually come to a halt. Kinetic energy is therefore a body's capacity to do work because of its velocity.

A body may also possess energy by virtue of the fact that it is in an elevated position. Potential energy is its capacity for doing work by virtue of its position. If a mass is lifted up to a certain distance the work done is said to be stored in the mass since we can get it back by letting the mass drop and this will be its potential energy. If the mass is allowed to fall freely it will lose its potential energy as it falls but this will be converted into kinetic energy as its speed increases. Just before it reaches the floor all its potential energy will have been converted into kinetic energy. In the next instant the body has reached the floor and has abruptly stopped. It has no potential energy or kinetic energy as the kinetic energy has been converted into two other forms of energy, heat and sound. If these were measured they would be exactly equal to the kinetic energy before impact. This is a simple example of a general principle known as the *Principle of the Conservation of Energy* which states that energy cannot be created or destroyed but can only be converted from one kind to another.

3 Machines

LEVERS

A MACHINE is a device which enables work to be done more easily and/or more quickly by applying forces. A lever is a simple machine which, when properly used, facilitates the task to be performed.

A lever is a straight bar or rigid structure of which one point, the fulcrum, is fixed, another is connected with the force or weight to be resisted or acted on, and a third is connected with the force or power applied. The force to be overcome is the load or resistance. The force applied to lift the load is the effort. (Figure 33)

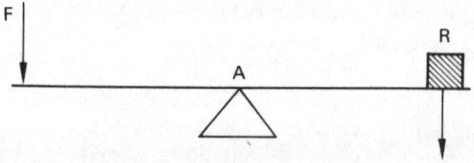

F = FORCE OR EFFORT
A = AXIS OR FULCRUM
R = RESISTANCE OR LOAD   FIG. 33

We use levers every day of our lives and they fall into three groups:

1   Take the examples of an old-fashioned can-opener, a crowbar, a wheelbarrow or a bottle-opener. Each of these can be represented by a rigid bar, and when a force is applied to one end of them it turns about a fixed point, the fulcrum, and overcomes a resistance. All these levers are used for the purpose of applying a relatively small force to overcome a relatively large resistance, and the range

of movement which the resistance covers is relatively small. (Figure 34.) In other words, the power to overcome a considerable resistance is gained at the expense of range of motion.

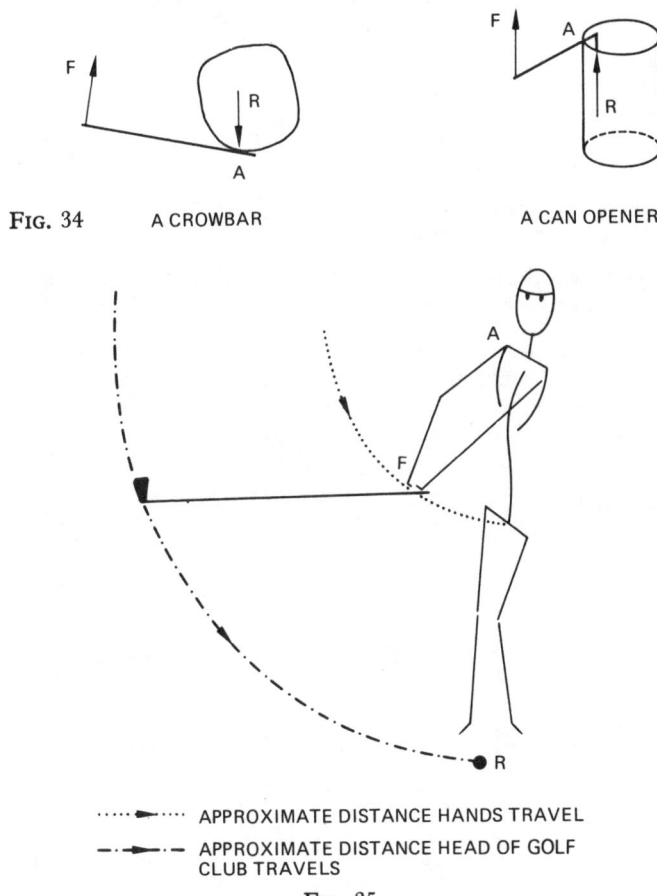

FIG. 34   A CROWBAR   A CAN OPENER

····▶···· APPROXIMATE DISTANCE HANDS TRAVEL

—·—▶—·— APPROXIMATE DISTANCE HEAD OF GOLF CLUB TRAVELS

FIG. 35

2   The striking implements used in sport are levers which do the opposite to the first examples. Tennis and squash rackets, baseball bats and hockey sticks are all types of levers used for the purpose of gaining distance at the expense of force. In a golf club the length of

the shaft enables the head to travel through a large arc of motion, but it is used to overcome the relatively slight resistance of the weight of the ball. By striking the golf ball with the club the player can impart more speed to it and send it a greater distance than he could by striking it with his hand, because the head of the club travels a greater distance and therefore at a greater speed than the hand. (Figure 35)

3   Levers such as a see-saw or scales gain neither force nor distance but provide for a balancing of weights. If the loads are equal they will balance when they are equidistant from the fulcrum. If they are unequal they will balance when the larger load is nearer to the fulcrum than the lighter load.

Levers are classified according to the relationship of the fulcrum, the force, and the resistance. There are three orders of levers:

1   A first order lever is one in which the axis lies between the point of application of the force and resistance points.
2   A second order lever is one in which the resistance point lies between the axis and the force points.
3   A third order lever is one in which the force point lies between the axis and the resistance points. (Figure 36)

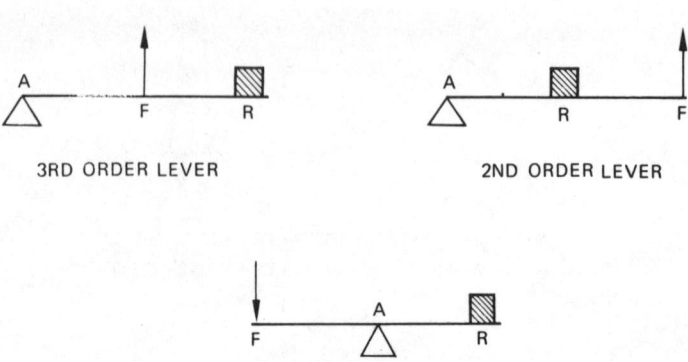

3RD ORDER LEVER   2ND ORDER LEVER

1ST ORDER LEVER

Fig. 36

Moment of Force

To understand the effect of forces acting on levers, the moment of force or torque must be discussed. In Figure 37 AB represents a rigid bar. At A, force F_1 acts down. At B, force F acts down. The

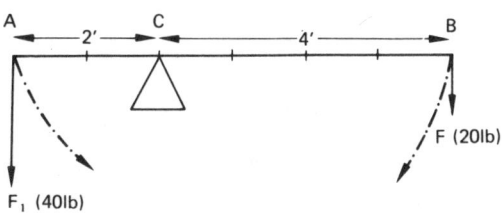

FIG. 37   1 cm = 1 ft.

bar can pivot about the fixed point, C. Force F is attempting to turn the bar AB about C in a clockwise direction. Force F_1 is attempting to turn the bar AB about C in an anti-clockwise direction. Experiments show that the effectiveness of any force to produce rotation depends on:

1 The magnitude of the force.
2 The length of the lever arm on which it acts.

The effectiveness of a force to produce rotation is called the *Moment of Force* or its *Torque* and is equal to the product of the force multiplied by the length of the lever arm on which it acts. In Figure 37 force F of 20lb. acting on the lever BC, which is four feet long, has a clockwise moment of force or torque of 20×4 which equals 80lb.ft. Equilibrium will be produced when the moment of force at A and B are equal but acting in opposite directions. Any number of parallel forces will be in equilibrium when the sum of all the clockwise torques are equal to the sum of all the anti-clockwise torques.

First Order Lever

In Figure 38 the force and resistance arms are equal and therefore the force and resistance must be equal. Should the force be applied at a greater distance from the fulcrum, the resistance must be placed nearer the fulcrum and vice versa. A first order lever is therefore

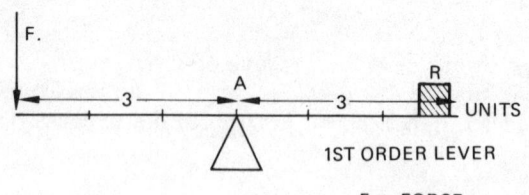

F.
A
R
3
3
UNITS
1ST ORDER LEVER
F = FORCE
R = RESISTANCE
A = AXIS

FIG. 38

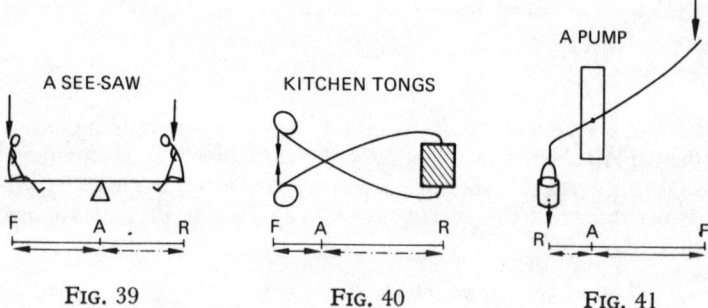

A SEE-SAW
F
A
R

FIG. 39

KITCHEN TONGS
F
A
R

FIG. 40

A PUMP
R
A
F

FIG. 41

F
R
A
R.A.
F.A.
2ND ORDER LEVER
A
R
F
RA = RESISTANCE ARM
FA = FORCE ARM

FIG. 42

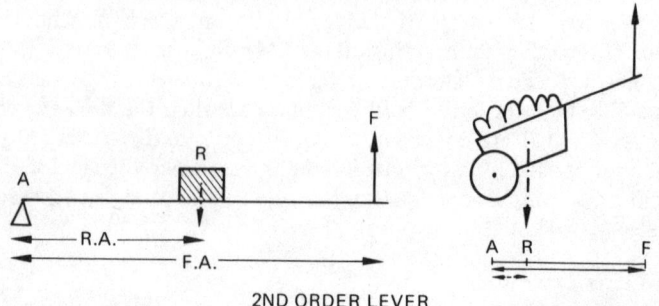

one of balance, in which the moment of force on each side of the fulcrum is the same. Figures 39, 40 and 41 are all examples of first order levers, but it can be seen that the relative length of the force arm and the resistance arm is different in each case.

Second Order Lever

From Figure 42 it can be seen that the force arm is the length of the whole lever and therefore in a second order lever the force arm is always longer than the resistance arm. A lever, whether first or second order, whose force arm is longer than its resistance arm, favours power because when the weight to be moved is nearer to the fulcrum than the effort, relatively less effort is required and a mechanical advantage is gained. This type of lever results in powerful movement which is often slow, e.g. a wheelbarrow.

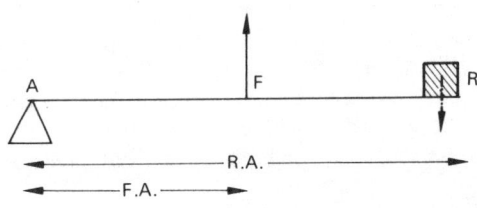

FIG. 43   3RD ORDER LEVER

Third Order Lever

From Figure 43 it can be seen that in a third order lever the resistance arm is the length of the lever. Therefore the resistance arm is always longer than the force arm. A lever, whether first or third order, which has a longer resistance arm, will favour speed, and distance, at the expense of force. An object of negligible weight can be moved more rapidly and to a greater distance by this kind of lever than without its aid, but the effort required will be great.

Principle of Levers

A lever of any order will balance when the product of force and the force arm equals the product of the resistance and the resistance arm.

$$F \times FA = R \times RA$$

Mechanical Advantage (M.A.)

The mechanical advantage of any machine is the relationship between the force arm and the resistance arm.

$$\text{M.A.} = \frac{\text{Force Arm}}{\text{Resistance Arm}}$$

Therefore, the greater the length of the force arm relative to the resistance arm the greater the mechanical advantage, that is, the less the force relative to the resistance needed to perform the task.

A first order lever will have a mechanical advantage of one, of more than one, or less than one, depending on the lengths of the resistance and force arms. A second order lever will always have a mechanical advantage of more than one as the force arm is always longer than the resistance arm. A third order lever will always have a mechanical advantage of less than one as the resistance arm is always longer than the force arm.

In the human body the muscles are applied through a system of levers. The levers are the bones, the fulcrum is the particular joint about which movement takes place, the force is the point of application of the specific muscle performing the movement and the resistance is the weight of the limb, or the limb plus a load.

Anatomical Examples of Levers

A good example of a first order lever is the skull in extension and flexion of the head. (Figure 44.) The skull is the lever, the atlanto occipital joint about which movement takes place is the fulcrum, the resistance is the weight of the skull whose centre of gravity lies anteriorly in the face, and the force is the extensor muscles of the neck acting from their attachment to the occipital bone. The action of this lever results in a state of balance, or movement, if the moment of force of the neck extensors is increased or decreased.

There are very few second order levers in the body. Some would argue that there are none at all. However, the action of brachio radialis muscle in flexing the elbow with the forearm in mid position is quite acceptable. The elbow joint acts as the axis about which movement takes place, the resistance is the weight of the forearm acting through its centre of gravity which lies at a point between the elbow and the radial attachment of brachio radialis which acts as the force. (Figure 45.) This is a lever of power and always has mechanical advantage.

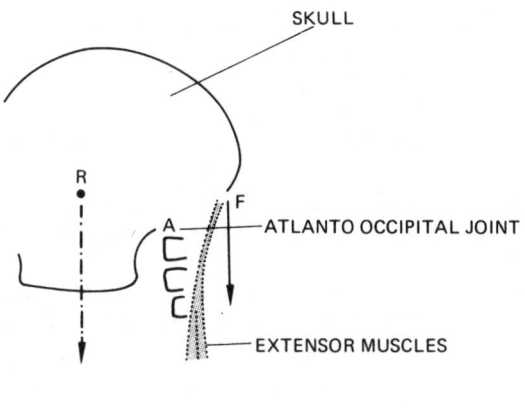

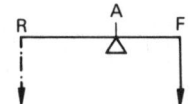

Fig. 44

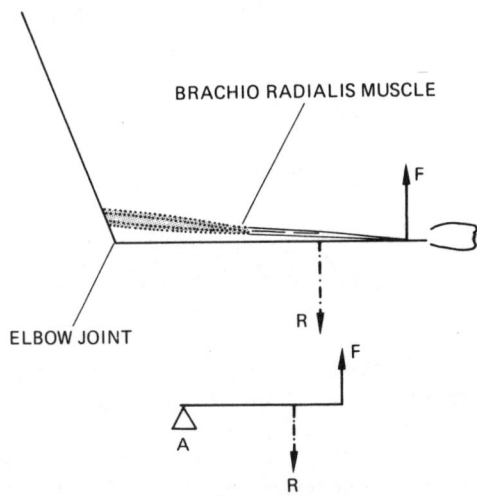

Fig. 45

Third order levers are the most common in the body. They always work under a mechanical advantage of less than one (i.e. a mechanical disadvantage) and therefore the force needed is great, compared with the resistance to be moved, but the advantages are ones of speed and range of movement. Deltoid acting on the shoulder to abduct the arm, brachialis acting on the elbow to flex the elbow, the hamstrings acting on the tibia and fibula to flex the knee are all examples of third order levers. In Figure 46 the fulcrum is the shoulder joint, the force is applied at the insertion of deltoid in the deltoid tubercle of the humerus and the resistance is the weight of the arm whose centre of gravity is distal to the force.

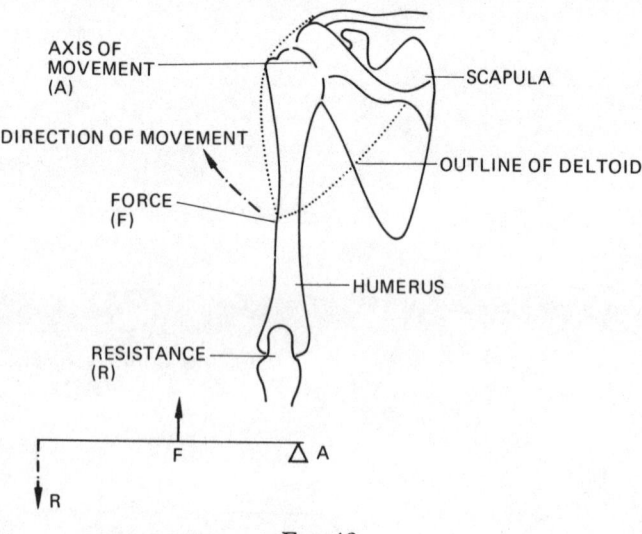

FIG. 46

In the body, the length of the force arm cannot be altered, as the force is applied at the attachment of a muscle, but the resistance can be changed by increasing or decreasing the length of the resistance arm or the actual resistance (deltoid acting with the arm flexed, or straight, or with a weight in the hand).

ANGLE OF PULL

Mechanically a force is most effective when it is applied at right angles to a lever. The direction of force exerted by a single muscle depends on the relationship of the moving bone's long axis to the muscle insertion. The angle between the muscle's line of pull and the long axis of this bone (the axis being a straight line between the mid points of both ends) is known as the angle of pull (Figure 47).

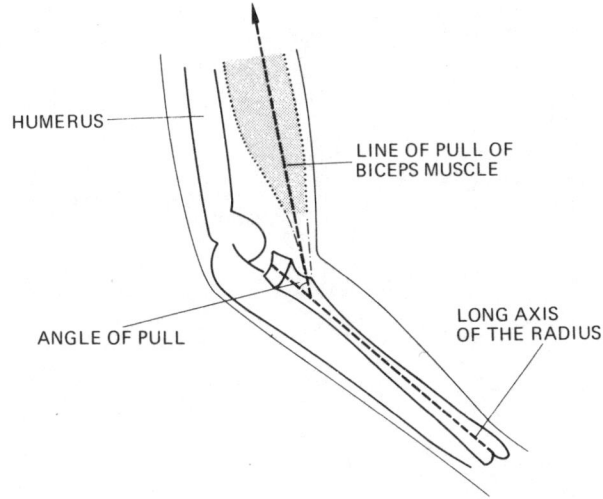

HUMERUS

LINE OF PULL OF
BICEPS MUSCLE

ANGLE OF PULL

LONG AXIS
OF THE RADIUS

FIG. 47

If the bone curves considerably the greater part of the axis will lie outside the shaft and so does not necessarily have to pass through the centre of the bony lever. The angle of pull is not fixed and changes with every degree of motion. Owing to the anatomical arrangement of muscles the angle of pull is liable to be small.

Should a muscle's line of pull be parallel with the bone's mechanical axis no motion would result. Therefore in order to produce movement the force must not only be sufficiently great to overcome the inertia of the part, but, in the case of levers, and therefore of body segments, it must be applied at an angle. There are no muscles in the body whose pull is parallel to the axis of the bone on which they act but there are muscles which pull at such a small angle that their importance in contributing to movement would appear to be negligible (Figure 48). These muscles pull the bones lengthwise

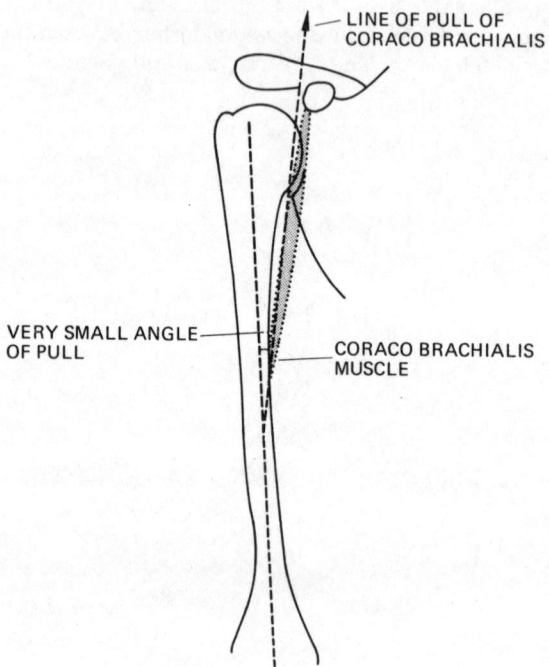

LINE OF PULL OF CORACO BRACHIALIS

VERY SMALL ANGLE OF PULL

CORACO BRACHIALIS MUSCLE

FIG. 48

towards their proximal joint and thus serve to stabilise them. Consider two muscle functions:

1   To produce movement.
2   To stabilise joints.

Except when a muscle is pulling at exactly 90° to the axis of the bone, the muscle force will have two components. The first will be a working or rotary component which will cause the lever to turn, and the second will be a non-working or non-rotary component which usually pulls in the direction of the proximal joint and is frequently called the stabilising component because it helps to

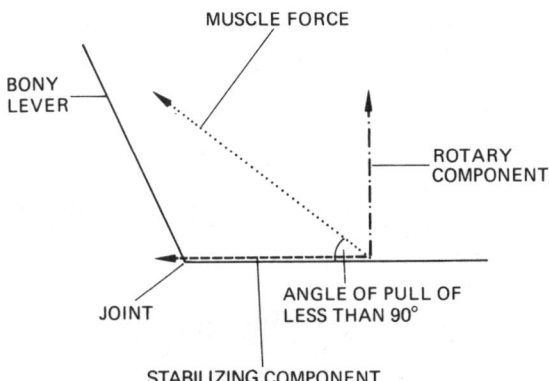

FIG. 49

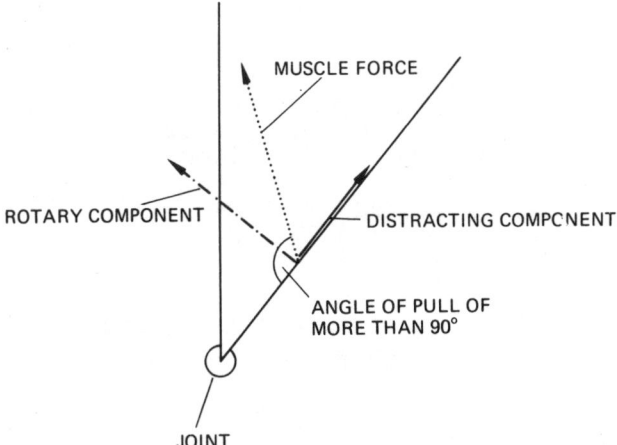

FIG. 50

stabilise the joint (Figure 49). However, when the muscle's angle of pull is greater than 90° the non-rotary component is pulling away from the proximal joint and is therefore a distracting component (Figure 50). This happens with few muscles and only when that muscle has contracted to its shortest position and therefore has little force left. There are very few muscles which have a large angle of pull. Most muscles are so arranged around a limb, that the axis of pull is almost parallel with the axis of the lever arm so that the angle of application is small. Therefore the stabilising component in most muscles is usually much larger than the rotary component. (Contrary to this is pectoralis major which inserts into the intertubercular sulcus at almost 90°.)

The Efficiency of a Resistance

The sustained pull of a force offering resistance will be maximal when applied at 90° to a lever and will decrease as the angle of pull becomes acute or obtuse. Resistance for the levers of the body would normally be given manually or using slings attached to a spring resistance, or with weights. When given manually, for instance to resist the flexors of the knee, one's hand can change its position of force so that it will always be at 90° to the leg (Figure 51). In Figure

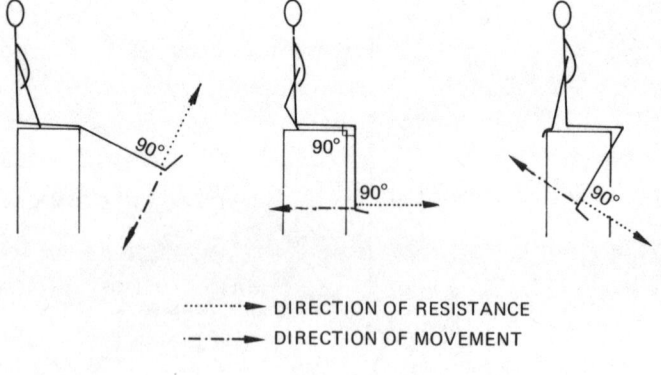

········▶ DIRECTION OF RESISTANCE
·—·—·▶ DIRECTION OF MOVEMENT

FIG. 51

51a the hamstrings will be acting at their most efficient as their angle of pull on the tibia and fibula is almost 90°. When resistance is given with graded springs the angle of pull will vary so the right-angled pull is employed in the range where maximal resistance is wanted. Re-education is usually most important in inner and outer range and therefore it is most important to have the resistance at 90° in these ranges. In graded exercises the resistance can either be increased by increasing the actual resistance or by increasing the length of the resistance arm thus increasing the leverage.

Resolution of Forces

A force is resolved, to determine its components. This is done when one wants to determine the components of a muscle force. A muscle force can be resolved into two components. One which acts at 90° to the bony lever and the other which acts along the lever to the proximal joint. In Figure 52 the vector F represents the tension

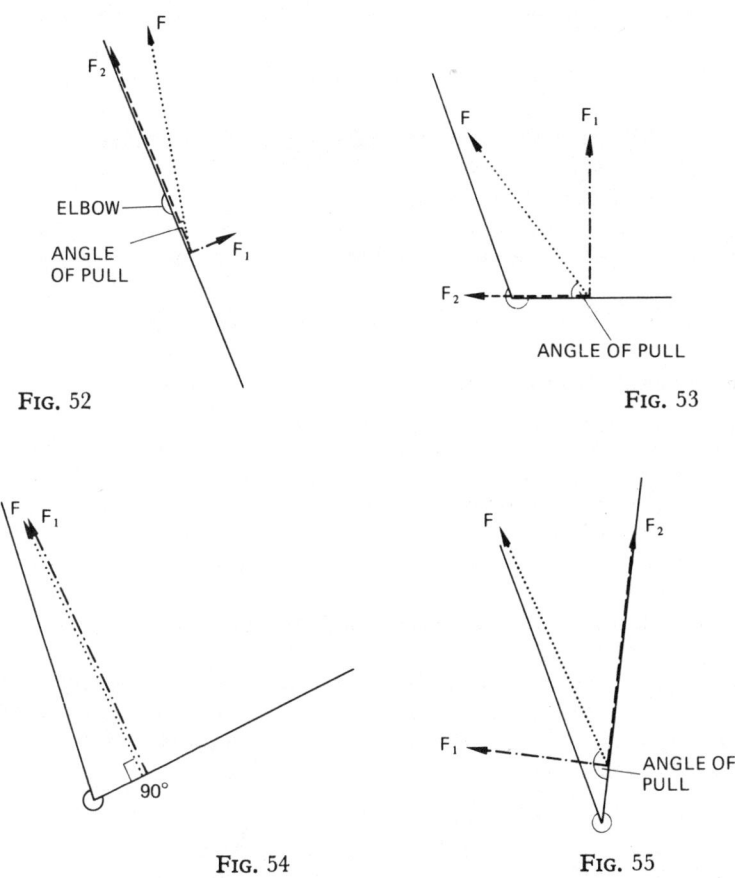

Fig. 52

Fig. 53

Fig. 54

Fig. 55

produced by a contracting muscle. F_1 is the rotary component acting perpendicularly to the bony lever and produces rotation at the joint.

F_2 acts along the long axis of the bone and presses the two joint surfaces together acting as a stabilising component. When joint movement takes place the magnitude of the two forces F_1 and F_2 change continuously as the angle of pull changes.

In Figure 52 the elbow is in extension and the pull of brachialis is represented by the vector F. The angle of pull is very small and therefore the rotary component F_1 is also very small. The stabilising component F_2 is large. As seen in Figure 53, as the elbow flexes, the angle of pull gets larger, the rotary component increases, and the stabilising component decreases. In Figure 54 the angle of pull is 90°. The rotary component is at its maximum and there is no stabilising component. As the angle of pull increases beyond 90° the rotary component decreases and the distracting component gets larger. (Figure 55)

If the angle of pull is known and the contractile force F is known then F_1 and F_2 can be obtained by using the formula:

$$F_1 = F \times \text{sine } \propto$$
$$F_2 = F \times \text{cosine } \propto \text{ where } \propto = \text{angle of pull.}$$

Anatomical Construction of Muscles

In terms of movement muscles are built primarily for power or speed. If one compares two muscles with the same volume but one having a few long fibres and the other having many short fibres, when they contract both will be capable of the same amount of work but they will have a different lifting power and lifting height. The muscle with a few long fibres will lift a lesser load a greater distance, while the muscle with many shorter fibres will have a lesser lifting height but will lift a greater load a smaller distance.

Long strap-like muscles with parallel fibres are built for speed and will therefore lift lesser loads a greater distance compared with muscles whose fibres are short, more numerous and pennate in structure, which are built for power and which will lift a greater load a smaller distance.

THE INCLINED PLANE

The inclined plane is the simplest of mechanical devices and is simply a slanting surface. In Figure 56 to lift the 50lb. weight above his head the man has had to exert a force of just over 50lb. Should this 50lb. weight be placed on the ground, the force needed to move it, would be that sufficient to overcome the friction between the load and the ground, once its inertia has been overcome, and this

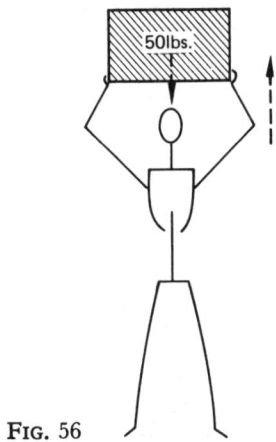

Fɪɢ. 56

would be very much less than a force of 50lb. as the ground is supporting the full weight. However, the weight has not been lifted through any height at all. An inclined plane is used to facilitate the lifting of a weight through a certain height.

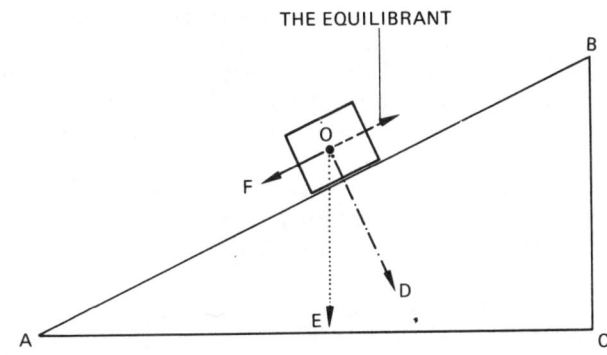

Fɪɢ. 57

In Figure 57, the object, whose centre of gravity acts at 0, is placed on an inclined plane AB. It is attracted by the earth and this is represented by the vector OE (this is the weight of the object). This force can be resolved into two components:

1 OF which tends to pull the object down the plane.
2 OD which tends to break the plane.

The force necessary to move the object up the plane to B, is equal and opposite to the force OF. (The equilibrant).

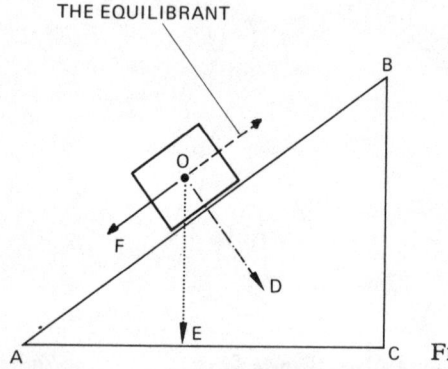

THE EQUILIBRANT

Fig. 58

Compare Figure 57 with Figure 58. In Figure 58 the inclined plane is obviously much steeper. The force OF is very much greater and therefore a greater force will be necessary to oppose this and move the object upwards.

The *Velocity Ratio* of a machine, by definition, is the distance travelled by a force, divided by the distance travelled by the resistance in the same time. In both Figures 57 and 58, CB is the effective distance the resistance has travelled. AB is the distance the force has travelled. It can be seen that the less the incline the greater the distance AB. Therefore as the plane declines the velocity ratio increases. Thus an inclined plane allows large loads to be lifted by allowing work to be done in low gear. The less inclined the plane, the longer it will take to achieve its objective but the less the effort that will be needed.

The uses of the inclined plane are many and varied. Ramps are used in football grounds, in theatres, hospitals and garages to make it easier to climb to a height (and easier to get down). Roads in hilly districts will prefer to zigzag rather than climb straight up.

A screw is an example of an inclined plane. If a triangular piece of paper with one of its angles at 90° is wound round a pencil, then a screw is produced (Figure 59). A screw is therefore an inclined

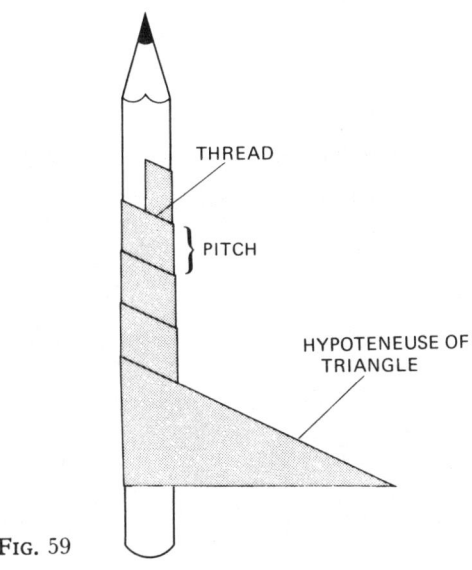

THREAD

} PITCH

HYPOTENEUSE OF TRIANGLE

FIG. 59

plane wound on a cylinder. The greater the number of threads, the less the incline, the higher the velocity ratio. The pitch of the screw is the distance between two threads. The velocity ratio equals the distance travelled by the effort divided by the distance travelled by the resistance and this equals $2 \pi R$/pitch of screw. Therefore the smaller the pitch the greater the velocity ratio, and the greater the mechanical advantage.

THE PULLEY

A pulley consists of a wheel which turns readily on an axle, which in turn is mounted in a frame (Figure 60). The wheel is usually grooved for a rope or a wire cable. The frame, or block, may be of

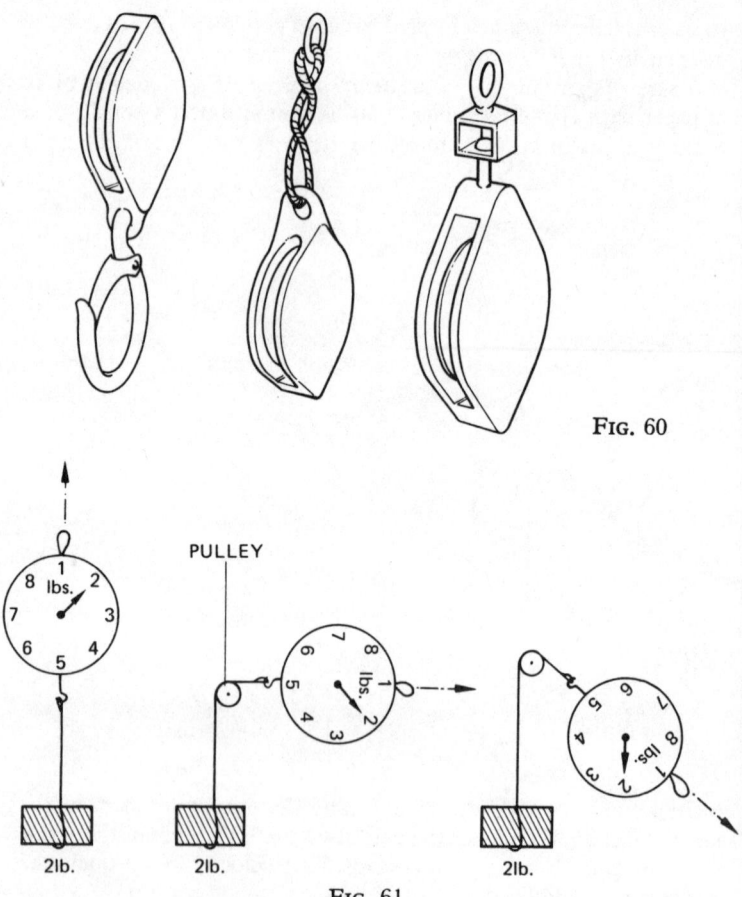

Fig. 60

PULLEY

2lb.   2lb.   2lb.

Fig. 61

metal or wood. More than one wheel may be mounted on the same frame, and more than one wheel may be mounted on the same axle. There are three types of pulleys to consider:

1 A single fixed pulley.
2 A single movable pulley.
3 Pulley combinations.

A single fixed pulley is a first class lever. It changes the direction but not the magnitude of a force acting on the pulley rope. The force remains the same on either side of the pulley rope irrespective of the angle of pull of the force (Figure 61). Figure 62 represents a single fixed pulley. The effort arm equals the resistance arm (both

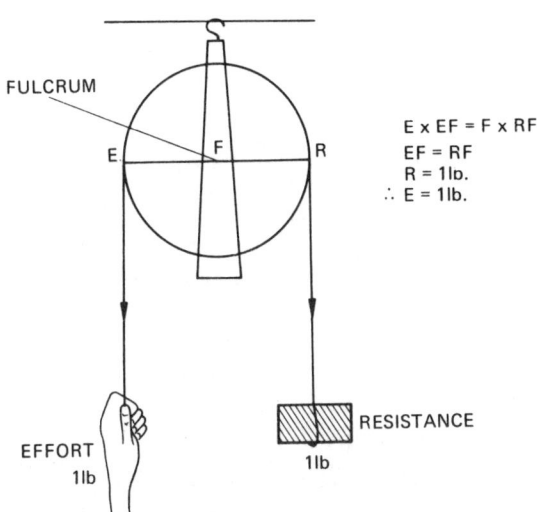

FULCRUM

E x EF = F x RF
EF = RF
R = 1lb.
∴ E = 1lb.

EFFORT
1lb

RESISTANCE
1lb

FIG. 62

are represented by the radius of the pulley). The mechanical advantage, EF/FR equals 1. Therefore when an effort of 1lb. pulls downwards through a distance of 1ft., a resisting weight of 1lb. will be raised 1ft. It gains neither force nor speed, but only changes direction.

A single fixed pulley is represented in the musculo-skeletal system to serve two purposes. It will either change the direction of the force to give a muscle a greater angle of pull, or it will change the direction of the force to produce a totally different movement to that which it would otherwise effect. The angle of pull of gracilis muscle as in Figure 63, is increased by means of the bulging medial condyles of the femur and tibia over which the tendon passes before it inserts into the tibia. In Figure 64 peroneus longus muscle passes behind the lateral malleolus before it turns under the foot to attach to the first cuneiform and the base of the first metatarsal, plantar

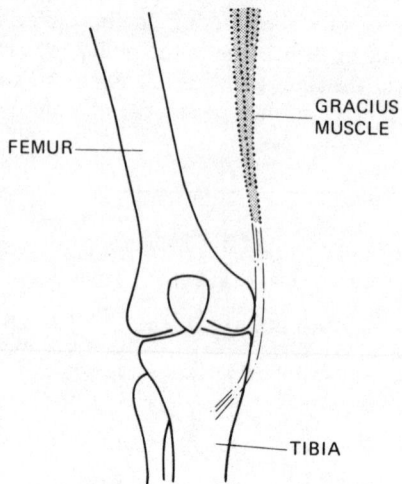

FEMUR

GRACIUS
MUSCLE

TIBIA

FIG. 63

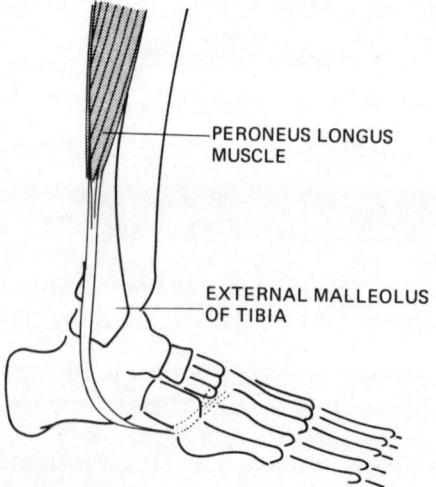

PERONEUS LONGUS
MUSCLE

EXTERNAL MALLEOLUS
OF TIBIA

FIG. 64

flexing the foot at the ankle joint. If it passed in front of the malleolus
its pull would be effective in front of the ankle joint and it would
therefore dorsiflex the ankle.

A single movable pulley acts as a second class lever. In Figure 65
the effort force E acts upon the arm EF which is the diameter of the

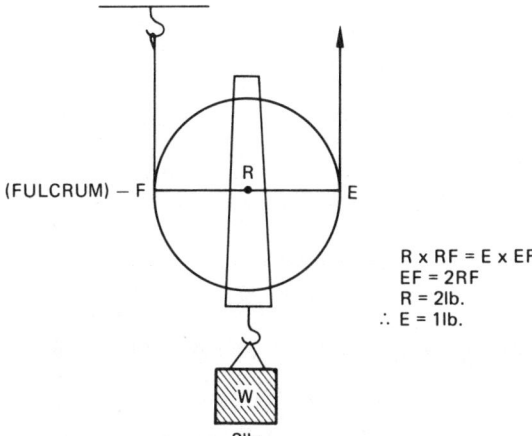

$R \times RF = E \times EF$
$EF = 2RF$
$R = 2lb.$
$\therefore E = 1lb.$

FIG. 65

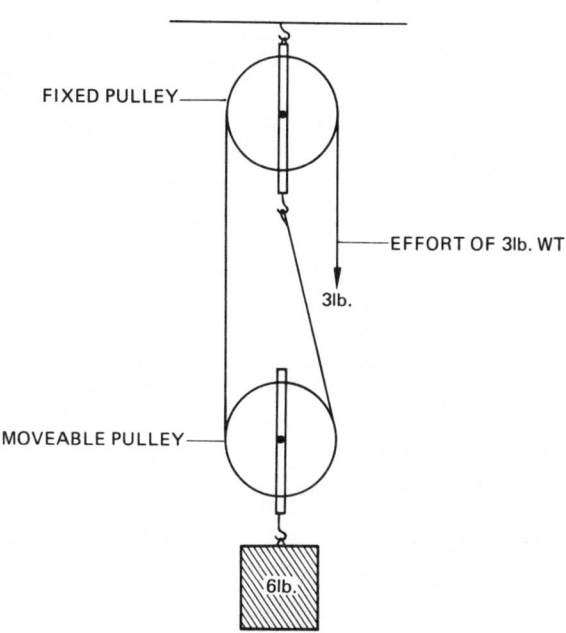

FIG. 66

pulley. The resistance (w) acts on the arm RF which is the radius
of the pulley. As EF = 2 RF, EF/RF equals 2. Therefore the
mechanical advantage of a single movable pulley is always 2. If both
ends of the cord are fixed, each cord must support half the weight
(i.e. 1lb. each). When the effort moves 2ft. the resistance will be
lifted 1ft.

Pulley combinations. There are many types of pulley combinations
using both fixed and movable pulleys. Figure 66 represents the
pulley combination used in the Guthrie-Smith suspension frame,
which is a combination of one fixed and one movable pulley. The
number of strands of rope supporting the movable pulley is two.
The last strand on which the effort force acts does not support it
but is used to change the direction of the effort force. The weight
is supported equally by each of the ropes leading to the lower pulley
block. Because the resistance weighs 6lb. there will be 3lb. tension
in each of the ropes. The mechanical advantage is the size of the
load divided by the size of the effort. Because there is a tension of
3lb. in each of the ropes, the effort is 3 lb. weight and therefore the
mechanical advantage is 6/3 which equals 2. This is also the number
of strands supporting the movable pulley. In any pulley combination
the mechanical advantage will be equal to the number of strands
supporting the movable block. (Figure 67). Figure 68 shows the
pulley combination used in physiotherapy departments. The upper
pulley is a single fixed pulley attached to an overhead support. One
end of the rope is fixed to a movable wooden bar, wound round the
fixed pulley and then round the movable pulley, up through the
wooden bar and then attached to the block of the fixed pulley. The
resistance in the form of a limb is attached to the movable pulley.
The effort is applied to the movable wooden bar. In the diagram
the weight of the limb is 10lb. The tension is the same in all parts
of the rope. The movable pulley is supported by two strands and
therefore the tension in each of these will be 5lb. The effort needed
will therefore also be 5lb. and the mechanical advantage is two.

If a machine were perfect then the work put into the system should
equal the work got out. In practice, however, some work is always
wasted in overcoming friction and in raising moving parts, and
therefore the useful work done by a machine is always less than the
work done by the effort.

The ratio of useful work done by the machine, to the total work put into the machine is the *Efficiency* of the machine.

$$\text{Efficiency} = \frac{\text{Work output}}{\text{Work input}} \times 100\%$$

(where Work = Force × Distance).

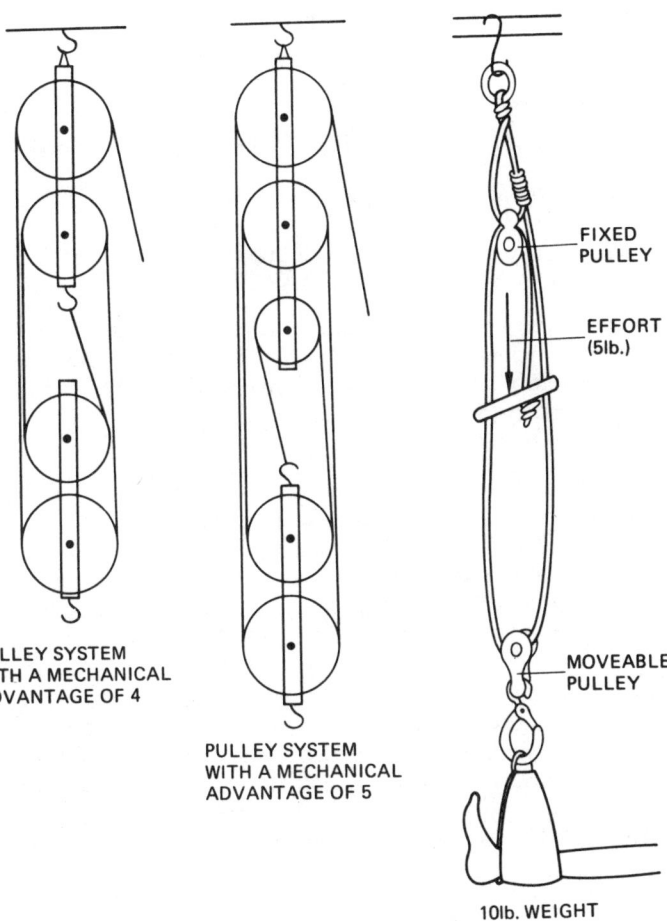

PULLEY SYSTEM
WITH A MECHANICAL
ADVANTAGE OF 4

PULLEY SYSTEM
WITH A MECHANICAL
ADVANTAGE OF 5

FIXED
PULLEY

EFFORT
(5lb.)

MOVEABLE
PULLEY

10lb. WEIGHT

FIG. 67   FIG. 68

SPRINGS

Elasticity is the property of a body which enables it to regain its original form after it has been distorted by the application of a force. The force which acts on a body and tends to distort it is called the stress. The change which is produced is called the strain. (E.g. the change in length per unit length.)

There are different kinds of elasticity:

a *Elasticity of Compression*, which can be illustrated by squeezing a rubber ball and distorting it. When the stress is removed, it returns to its original state.

b *Elasticity of Extension*, which can be illustrated by the stretching of an elastic band.

c *Elasticity of Torsion*, which can be illustrated by the twisting of a coiled spring.

d *Elasticity of Flexion*, which can be illustrated by the bending of a strip of steel.

In all these cases the material tends to resume its original form when the stress is removed. The strain produced in these materials will be as a result of two basic kinds of change. Either the change will be an alteration in the volume, or the shape of the material (or a combination of both).

Hooke's Law states that the strain is proportional to the stress producing it (but this law holds for small strains only). This can be demonstrated by fastening one end of a steel wire to a beam. Weights are then added to a hook on the end of the wire, and it is found that for the addition of a weight of X gm. the wire will stretch 1mm. and when the weight is taken off the wire reverts to its original length. When a weight of 2X gm. is added, the wire will stretch 2mm. but again, the wire will revert to its original length when the weights are taken off. Eventually as an increasing number of weights are added a point is reached where the wire readily stretches a short distance with only a slight addition to the weights. This is its *Elastic Limit*. When all the weights are removed, the wire remains distorted and does not return to its original length; it has acquired a *Permanent Strain* or *Set*. If more weights yet are added there will come a point when the wire snaps; the *Breaking-point* of the wire has been reached. Therefore, Hooke's Law is applicable only until the elastic limit is reached.

The value of the ratios stress : strain is different for different materials, but is reasonably the same for the same material even though it is fashioned in different shapes and sizes. This ratio is a means of comparing the elasticity of different materials and is called the *Elastic Modulus*. The greater the stress needed to produce unit strain, the higher the elastic modulus of a material.

All solids possess the property of elasticity to a certain degree, though for most it is not noticeable. The only metal in which elasticity is very well developed is steel, and then it needs to be hardened steel. Hardened steel is steel which on heating to red heat is then cooled very rapidly in cold water. This is a process called tempering. Tempered steel is hard and brittle but very elastic. To make it less brittle it is gently reheated. It looses some of its elasticity in the process but it is much softer and tougher. In a physiotherapy department all springs will be made of hardened steel. The elastic property of many substances is made use of. There are many examples including sorbo rubber, rubber balls, polyurethane, and piano wire to name but a few.

A spring consists of a uniform coil of wire which is extensible. When one end of the spring is fixed, it can be elongated by a force applied at the other end of it in the direction of its long axis. The increase in length of the spring is directly proportional to the magnitude of the force applied to stretch it. Let us consider two types of springs used for therapeutic purposes:

1   Short tension springs.
2   Long spiral springs.

Short Tension Springs are about 4 inches long and vary in tension according to their function. These are used in suspension therapy to allow a small amount of vertical movement and thus impart a feeling of buoyancy to the part suspended.

Long Spiral Springs or helical springs average about 12 inches in length. The springy metal is relatively soft and it yields easily. Each spring is fitted at either end with a split clip which is capable of rotating in the tapered end of the spring to allow rotatory adjustment without distortion. The poundage of each spring is stated on metal tabs affixed to the split clips, and this will be the force needed to stretch the spring to its maximum. This is indicated by a tape within the coil which is taut when the maximum pull is reached. The tape will also prevent the coil from being overstretched. Spiral springs are used as resistances to movement, as assistances to movement, when the recoil of the spring will be used, and as oscillators.

Two or more springs can be used in series or in parallel. The force needed to stretch springs in parallel to maximum stretch will be the sum of the strengths of the individual springs. In Figure 69,

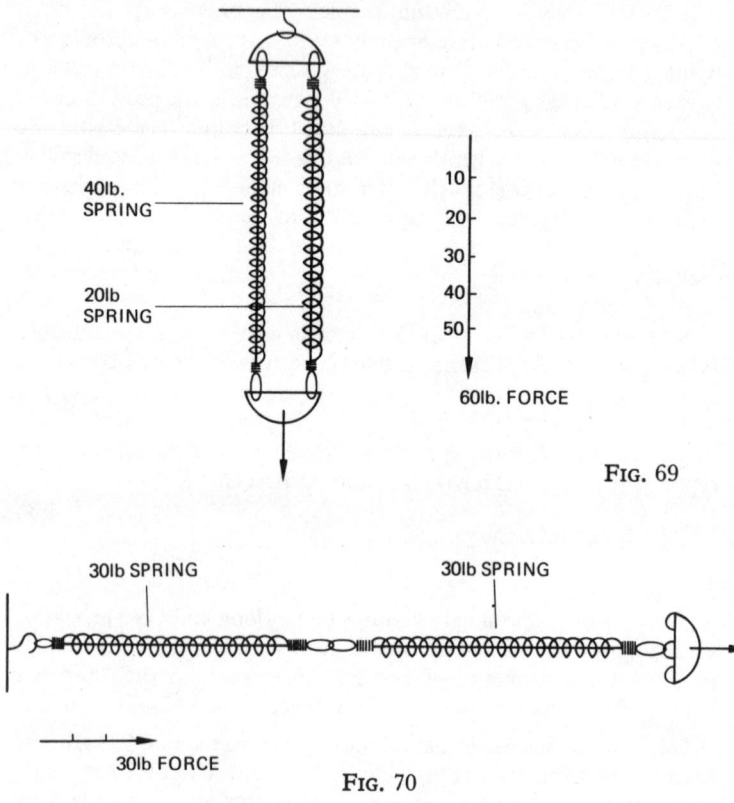

40lb. SPRING

20lb SPRING

10
20
30
40
50

60lb. FORCE

FIG. 69

30lb SPRING   30lb SPRING

30lb FORCE

FIG. 70

two springs are set up in parallel. The force needed to pull these two springs to maximum length will be (40 + 20)lb. which equals 60lb. They will therefore have the same effect as one 60lb. spring. When two equal springs are in series, the force needed to pull both springs out to maximum length will be the same as for just one of the springs. In Figure 70 two 30lb. springs are placed in series. A force of 30lb. is needed to extend both springs completely, but the whole arrangement must be extended to twice the length needed for the springs in parallel to get maximum pull.

THE PENDULUM

A pendulum is a body suspended in such a manner that it can swing to and fro about a horizontal axis. A simple pendulum will consist of a small heavy bob, suspended by a light thread from a fixed support (Figure 71). If the bob is displaced through an angle

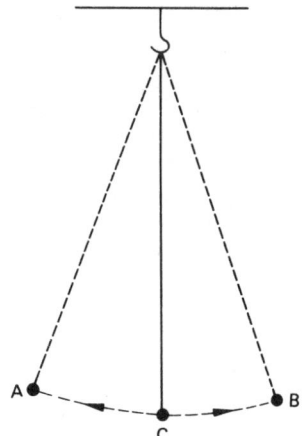

Fig. 71

of not more than 20° and then released, it will swing to and fro at a regular frequency. (Frequency is the number of cycles per second.) One complete oscillation or cycle will be from C to A, from A to B and then back to C. The time taken for one complete oscillation is called the Periodic Time. The greater the time taken to complete one oscillation, the less the number of oscillations per second, and vice versa, i.e. the frequency is inversely proportional to the periodic time.

Therefore $T = 1/n$

where T = periodic time in seconds and n = frequency.

The longer the pendulum the greater the periodic time and therefore the slower the frequency. In a simple pendulum it is found that T is proportional to the square root of its length, so if there are two pendulums, one 25 cm. long and the other 100 cm. long, the periodic time T of the longer one will be twice that of the shorter one.

The limbs of the body are complex pendulums. If the leg is locked at the knee, and swung to and fro from the hip, its natural frequency of oscillation can be found roughly, if care is taken not to force the swing with the muscles. The same can be done for the lower leg swinging from the knee. Its natural frequency will be considerably less than that of the swing from the hip (1·9 and 1·1 seconds respectively). In comfortable slow walking with the minimum of muscular effort, the periodic time of the stride is roughly equal to the period of free oscillation of the whole leg about the hip. If the stride is more rapid, then the legs are forced to swing faster than their natural frequency, and more muscle effort is required. In running the pendulum is shortened by bending the knee during the forward swing. (Figure 72). On the other hand, the length of the pendulum can be increased by wearing heavy boots, which lowers the centre of gravity and therefore lengthens the arm of the pendulum. Hence the slow rhythmical stride of the hill walker who is more or less carried along by the inertia of his boots!

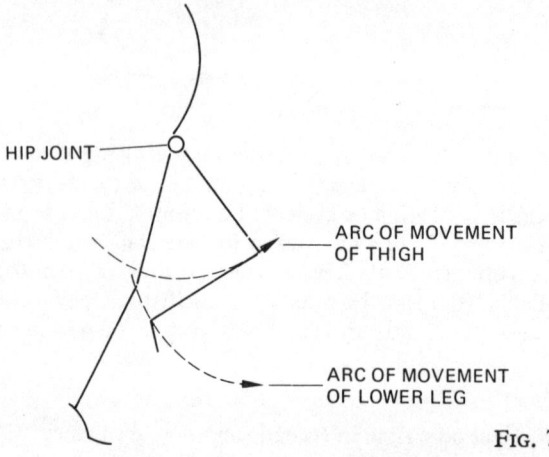

HIP JOINT

ARC OF MOVEMENT
OF THIGH

ARC OF MOVEMENT
OF LOWER LEG

FIG. 72

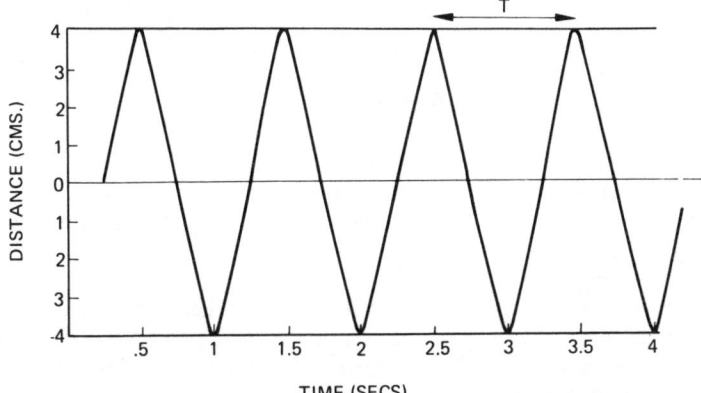

FIG. 73

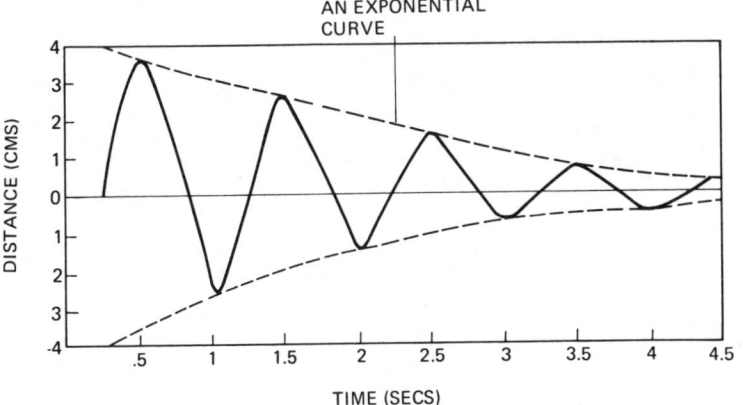

FIG. 74

Figure 73 shows a graph of displacement against time, for an undamped pendulum, which produces a continuous sustained oscillation of constant amplitude. However, in practice if a pendulum is left swinging, the amplitude gradually decreases as energy is lost by friction to the surrounding air. The oscillation is then said to be damp and unsustained. (Figure 74). In a damped oscillation the amplitude of each swing is less than the preceding swing by a constant factor which depends on friction. In practice all mechanical oscillations are damped unless energy is continuously supplied to replace losses caused by friction.

Resonance

If a pendulum is pushed, it is set in motion and will continue to swing as long as the push is continued. If the pushes are timed to equal the natural frequency of the oscillation, the pendulum oscillates with a large amplitude. If the push is at too high or too low a frequency the amplitude is much less. The hand represents the driving force and the pendulum is an oscillatory system, and when the frequency of the driving force equals the natural frequency of the oscillatory system, the amplitude is maximum and the two are said to be in resonance with each other.

Pendular movements are made use of in the Guthrie-Smith suspension apparatus. There are two different types of fixation:

1 Vertical or pendular.
2 Axial.

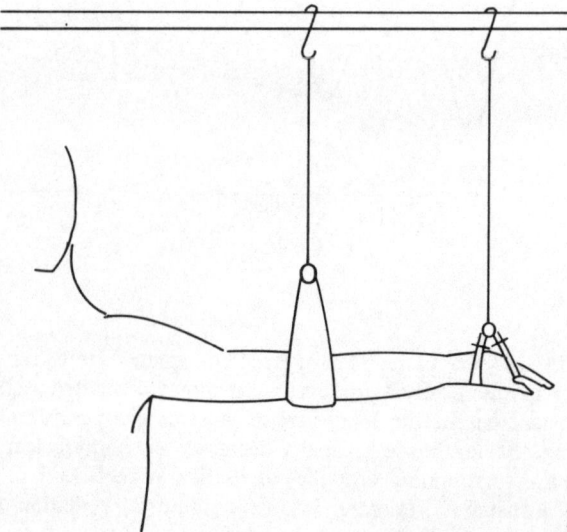

FIG. 75

Vertical or pendular fixation. In this form the limb to be moved is supported by slings, and each part is supported by a vertical pulley to the frame above (Figure 75). The limb is in a state of equilibrium as the weight of the limb acting down is equal and opposite to the uplifting force of the sling and pulley combination. If an initiating force is applied in one direction, the limb acts as a pendulum and will move upwards and away from the force, the amplitude of the swing being dependant on the length of the pulley rope and the magnitude of the force applied. The limb will continue to move upwards and out, until the force applied is equalled by the force of gravity. The limb will then fall down due to its potential energy and return to its original state of equilibrium. However, it now has kinetic energy which carries it outwards and up in the opposite direction (Figure 76). This alternating movement oscillates about the point of equilibrium and will continue but diminish, and eventually stop when the initial force is expended in overcoming friction and gravity. The movement can be prolonged if an additional force is applied in the direction of the movement. It can be halted if a force is applied in the opposite direction.

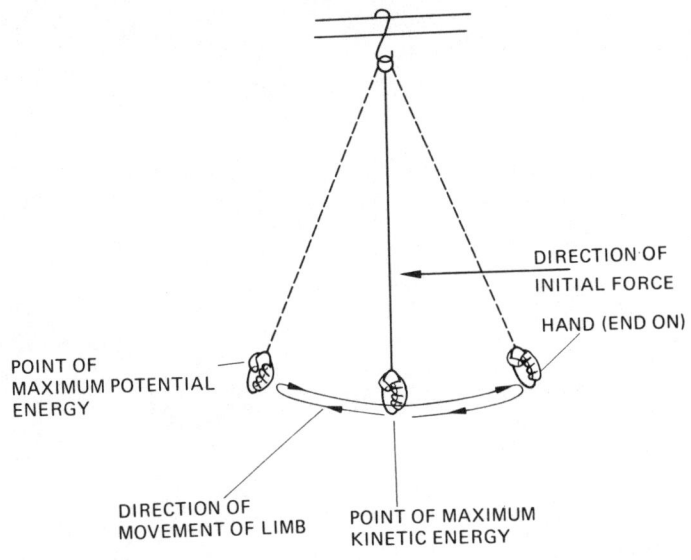

FIG. 76

Axial or balanced fixation. In this form all parts of the limb or limbs are suspended immediately above the axis of the joint to be moved. The movement is no longer pendular, but occurs in a single horizontal plane, and gravity will not exert a pull on any part of the movement (Figure 77), and therefore another force must be applied to maintain movement. This type of suspension is used where an increased range of movement is aimed for, because the movement is not limited by the length of the pulley as in pendular fixation. The assistance or resistance to the movement will depend on the point of fixation.

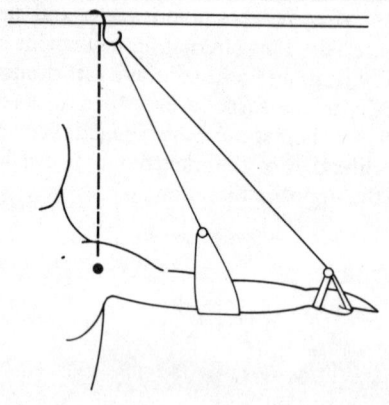

Fig. 77

4 Liquids

ANYTHING which occupies space is matter, and matter exists in the three forms of solid, liquid and gas. Water is an example of a liquid which can easily exist in all three forms, i.e. ice, water and steam. In common with other forms of matter, water will have certain physical properties of:

Mass
Weight
Density
Specific gravity
Buoyancy
Surface tension
Refraction
Viscosity
Hydrostatic pressure

MASS is the amount of material, i.e. the molecules and atoms, it consists of.

WEIGHT is the force with which a substance is attracted towards the centre of the earth, and is therefore the effect of gravity on its mass.

DENSITY is the relationship between mass and volume. If exactly equal volumes of different substances are weighed, such as feathers, balsa wood, and lead, it will be found that their weights are very different, and therefore their masses must be different. To compare these masses in a given volume is to compare their densities. The density of a substance is the mass of unit volume of that substance.

E.g. the density of gold is 19·3gm./cc.

Therefore 2gm. of gold would have a mass of 38·6gm. etc.

Thus density = mass/volume = 38·6/2 = 19·3gm./cc.

Water is at its most dense at 4°C. It expands at both lower and higher temperatures. Therefore ice will be less dense than water and for this reason it will float. (Figure 78.) The density of water is

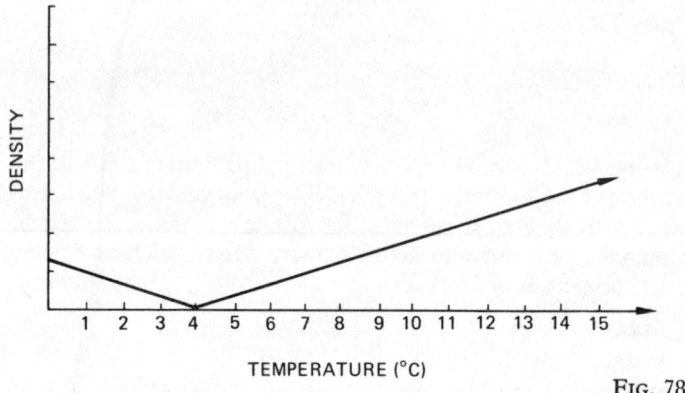

TEMPERATURE (°C)

FIG. 78

1gm./cc. Dissolved substances increase the density of water and therefore sea water is denser (1·024gm./cc) than fresh water (1gm./cc). The following are examples of the density of various substances:

$$
\begin{aligned}
\text{Ice} &- 0·92\text{gms./cc} \\
\text{Human body} &= 0·95\text{gms./cc} \\
\text{Iron} &= 7·7\text{gms./cc} \\
\text{Wood} &= 0·75\text{gms./cc}
\end{aligned}
$$

SPECIFIC GRAVITY, or relative density, is the density of the material compared with that of water (which is 1gm./cc.). Therefore, the specific gravity of a substance such as gold, which has a density of 19·3gm./cc, will have a specific gravity of 19·3. A body with a specific gravity of less than 1 will float in water, and a body with a specific gravity of more than 1 will sink in water.

HYDROSTATIC PRESSURE. A liquid such as water pushes at the sides as well as on the bottom of a container in which it rests. It will also press upwards on anything placed in it. The pressure at any point in a free-standing liquid is directly proportional to the depth below the surface. In Figure 79 the container has a cross

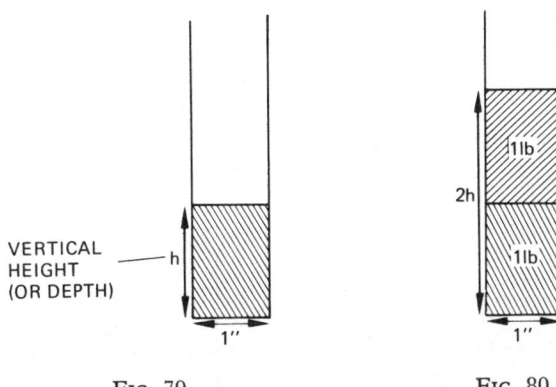

FIG. 79   FIG. 80

sectional area of 1sq.in. It contains 1lb. of water. The pressure on
the bottom will be Force/Area $= 1/1 = 1$lb./sq.in. In Figure 80 the
container is the same but it now holds 2lb. of water. The pressure
on the bottom will therefore be 2lb./sq.in. Therefore the pressure
is directly proportional to the depth below the surface. The depth
referred to is always the vertical depth. At any point within a liquid
that is at rest at the same vertical depth the pressure is the same in
all directions. So at any point, a liquid exerts an equal pressure in
all directions, even pushing upwards on the bottom of an object
immersed in it.

BUOYANCY. In Figure 81 a rectangular object is hanging in water.
The pressure on all its sides cancel themselves out. Since pressure
increases with depth, the upward force on the bottom of the object
will be greater than the downward force on the top. Therefore there
is a net lifting force. The object is lighter in water than it would be
in air and this lifting force is called *Buoyancy*.

In order to float, an object must be able to push aside its own
weight of water. Consider a 'block' of water in water (Figure 82).
This block floats because its downward pressure is balanced by an
equal upward pressure. If this block is removed the forces that had
been holding it up are now available to support some other object
of the same weight or less. Water has a specific gravity of 1. Therefore
an object with a specific gravity of 1 or less will be fully supported
by the water. The smaller the specific gravity of an object the more
it will float. (I.e. the greater its volume above the water-line.)

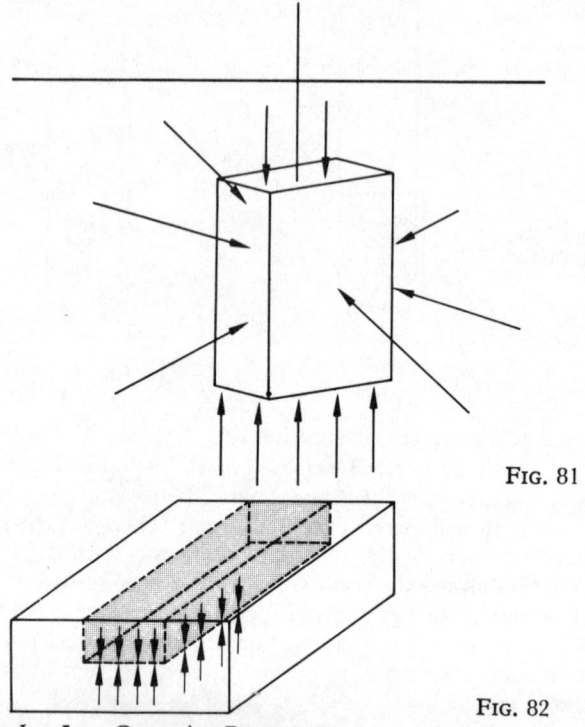

FIG. 81

FIG. 82

Archimedes: Law Governing Buoyancy

Any object immersed in a liquid appears to lose an amount of weight equal to that of the liquid it displaces.

If a body is able to float in water it means that the buoyant force is equal to the whole weight of the body. The body therefore appears to have lost its entire weight. When the lungs are filled with air, the human body has a slightly smaller overall density than water, and therefore it will float. However, it will be almost completely immersed, in order to displace the amount of water equal to the weight of the body. The greater the air volume in the lungs the easier it will be to float.

Objects, even if they are denser than water and sink, appear to lose some of their weight when submerged. A man can lift a larger stone under water than he can possibly lift in air. The upward force of the water lifts part of the weight for him.

If an object is weighed by hanging it on a spring balance and then is lowered into water, the reading falls as more and more of the object is immersed. Once the object is completely immersed, further loss of weight does not occur. The water has exerted an upthrust on the object, and this upthrust is equal to the weight of water displaced by the object. Thus if the object weighs 30gm. in air and on being lowered into water the spring balance reads 25gm., then the 30gm. weight must have displaced 5 gm. of water. If this is done in a different way and the water displaced is collected (Figure 83) and found to be 5gm., then the weight of the object in water must be 25gm. This will also give the volume and thus the density of the object. The upthrust of 5gm. weight is the weight of water displaced. 5gm. of water occupies 5cc. Density = Mass/Volume = 30/5 = 6gm./cc.

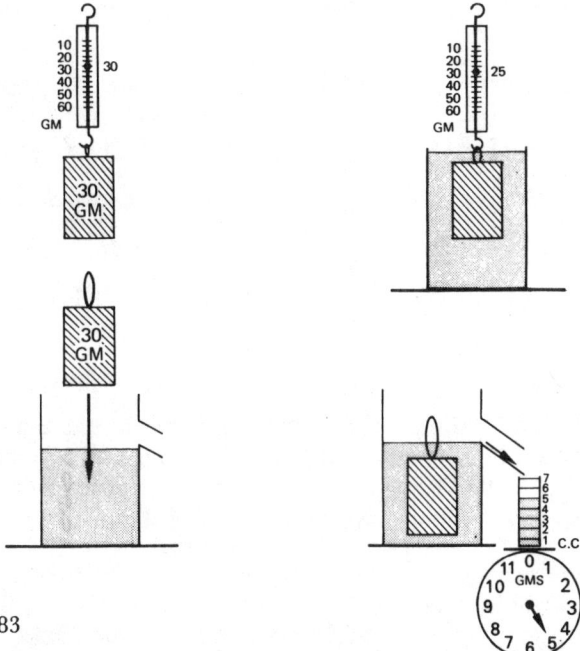

Fig. 83

When an object floats, it appears to lose all its weight as the water completely supports it. However, some of the object will lie under the water line, and it is therefore displacing some water. This, according to Archimedes' Principle, must be equal to its own weight. (The upthrust on the floating object must be equal to its weight and therefore the object must displace its own weight of water.)

A body in water is subjected to two opposing forces:

1   Gravity, which acts downwards at the centre of gravity of that body.
2   Buoyancy, which acts upwards through the centre of buoyancy, which is the centre of gravity of the displaced water.

These will be equal and opposite as the weight of the floating body is equal to its water displacement.

When the weight of a floating body equals the weight of the liquid displaced, and the centre of gravity and the centre of buoyancy are in the same vertical line, then the body is in equilibrium. (Figure 84a).

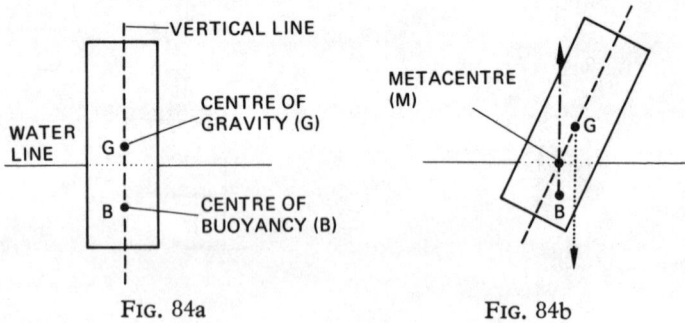

FIG. 84a   FIG. 84b

If the body is tipped, the centre of gravity and the centre of buoyancy will no longer be in the same vertical line. The centre of gravity will not change, but the centre of buoyancy will alter according to the position the body is tipped into. (Figure 84b). These two forces now form a torque, and this torque will either restore the body to its original position or will urge the body further from its original position. Whether a body will float stably or not will depend on a certain point in the body called the *Metacentre*. The metacentre is the point through which the force of buoyancy acts in both stable and unstable positions. In Figure 84a, the force of buoyancy acts upwards through the vertical line. In Figure 84b, the force of buoyancy acts through the metacentre, which is the point at which the vertical line from the centre of buoyancy meets the original vertical line running through both the centre of gravity and centre of buoyancy.

A body will float stably when its metacentre lies above its centre of gravity (Figure 85), and will be unstable when its metacentre lies below its centre of gravity (Figure 84b).

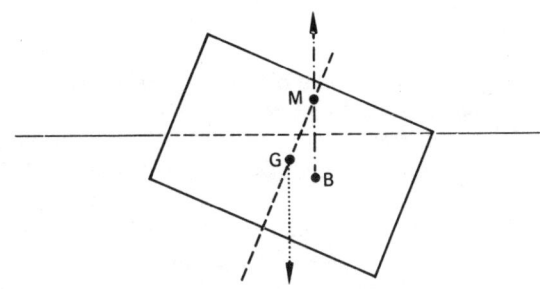

FIG. 85

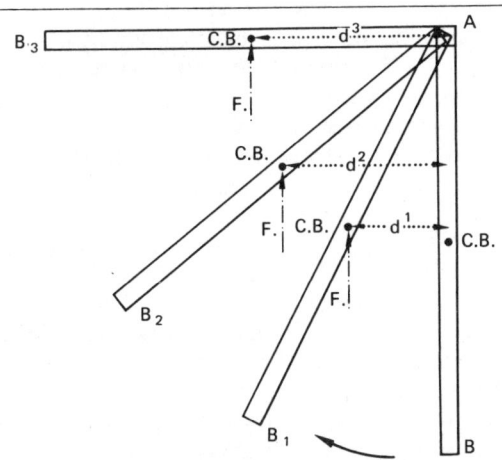

FIG. 86

The moment of force about a point is the turning effect of the force about that point. Buoyancy is itself a force, and is therefore governed by this rule. The moment of buoyancy is the force of buoyancy, multiplied by the perpendicular distance from a vertical line from the point about which the turning effect of buoyancy is exerted, to the centre of buoyancy. In Figure 86, AB represents a lever submerged in water, A representing the point about which the turning effect of buoyancy is exerted. F is the force of buoyancy, CB is the centre of buoyancy, and d is the perpendicular distance from the vertical line. The moment of buoyancy on $AB_1 = F \times d^1$, on $AB_2 = F \times d^2$, and on $AB_3 = F \times d^3$. When the lever is vertical (AB), $d = 0$ and therefore there is no moment of buoyancy and therefore there is no turning. At AB_2, d^2 is greater than d^1 and therefore the moment of force is greater on AB_2 than AB_1. The maximum moment of force is at AB_3.

In the human body the limbs act as the levers. In Figure 86, A can represent the joint about which movement occurs. If A represents the shoulder joint, it can be seen that the effect of buoyancy increases with abduction at the shoulder and is at its maximum when the arm lies parallel to, and just under the surface of the water. If the lever is shortened by bending the elbow the centre of buoyancy will move nearer to A. The distance, d, will be shortened, and therefore the moment of buoyancy will be less.

Buoyancy can be used to assist movement when the limb will be taken towards the surface of the water, and to resist movement when the limb is taken from the surface to the vertical. The moment of buoyancy increases as the limb moves nearer to the surface of the water and when the lever is lengthened. Therefore when re-educating weak muscles in water it is preferable to make use of as long a lever as possible, if maximum assistance or resistance is required.

COHESION and ADHESION. Cohesion is the force of attraction between neighbouring molecules of the same type of matter. Adhesion is the force of attraction between neighbouring molecules of different types of matter. Any solid or liquid is bound together by strong attractive or cohesive forces between the molecules. These act only over a very short distance, so a molecule will only influence the molecules immediately around it. When a molecule is in the middle of a liquid, it is attracted fairly equally by molecules on all sides, so that all the forces acting on it will tend to cancel each other out. If water is placed in a glass container the molecules next to the glass will be attracted:

1 in towards the rest of the liquid by the cohesive force of the surrounding water molecules and
2 away from the rest of the liquid by the adhesive force exerted by the glass molecules.

There will also be an adhesive force between the water and the air above, but this will be very weak as the relative concentration of air to water molecules is very small. Because the cohesive force between the molecules of water and the adhesive force between the water and the container, is greater than the adhesive force between the water and the air above, the molecules of water are drawn downwards and the surface of the water in the container is concave. Because the adhesive force between the container and the water is greater than the cohesive force of the water, when the water is poured out of the container the walls of the container will remain wet.

In a small bore tube placed in a beaker of water the adhesive force between the water and the glass is greater than the cohesive force of the water, and so the water molecules are attracted up the tube by the adhesive force. Nearby water molecules will follow because of the strong cohesive forces between them. This process continues, filling up the space below the surface as the water is pulled higher and higher. It continues until the upward attractive pull is equal to the downward pull of gravity acting on the column of water. This is called capillary action and occurs when tubes of fine bore are placed on end in water. The narrower the bore the greater the rise, as the greater the glass/water ratio will be. The adhesive force between the glass and the water produces the meniscus.

SURFACE TENSION. In an expanse of water, cohesive forces will hold the molecules of water together. On the surface, however, the upward attraction of water and air molecules is negligibly small, and this does not balance the downward force of the molecules below. This downward pull will tend to pull the surface down but the surface molecules cannot move because all available spaces below are already filled, so the surface is like a stretched piece of elastic in a continual state of tension, and this is called the surface tension.

Surface tension is the tendency of a liquid surface to contract as if it were covered by a tightly stretched membrane. In Figure 87, molecule A is attracted equally in all directions by the cohesion of surrounding molecules. Molecule B is attracted sideways and downwards but not so strongly upwards. Molecule C is only attracted upwards by the weak adhesive force between it and the air molecules above and there is therefore an unbalanced force which tends to pull the molecule downwards towards the interior of the liquid. This force causes the surface to act like an elastic film.

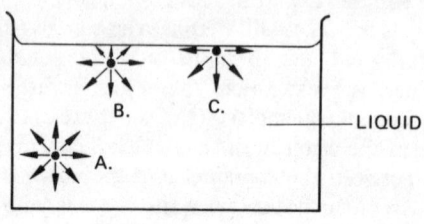

FIG. 87

REFRACTION. Refraction is the bending of a ray of light from one medium to another of a different density. (E.g. from water to air.) When a ray passes from a rarer to a denser medium, the ray will be bent towards the normal, the normal being a line at 90° to

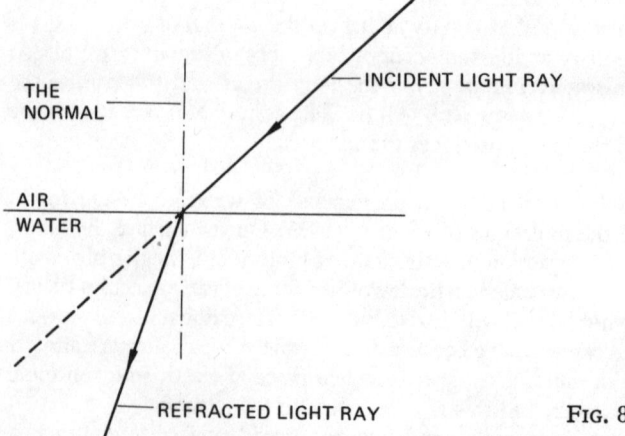

FIG. 88

the dividing line between the two mediums. Figure 88 shows a ray of light passing from air to water where it is refracted towards the normal. If that ray of light were to pass in the opposite direction (i.e. from water to air) it would be refracted away from the normal.

If there is an object at the bottom of the water, and a ray of light is reflected on it, then the rays would be refracted away from the normal as they pass from water into air. (Figure 89.) The eye

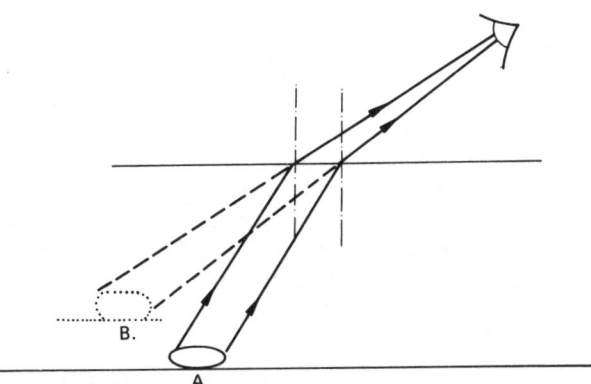

FIG. 89

presumes that light travels only in straight lines and sees the object A at B. The water seems to be more shallow than it really is and the object appears to be at a lesser depth. This is an optical illusion. Because of this it is very difficult to judge accurately the movement of limbs in water.

VISCOSITY. Viscosity is due to internal resistance within a liquid and is caused by friction between the molecules of that liquid. Friction is an obvious property between two solids (e.g. a brick on a rough surface, where the force of friction between the two surfaces will oppose movement). The frictional forces are much weaker in liquids but are present in all liquids and force is needed to overcome them. Gases are even less viscous than liquids but some will be more viscous than others. (E.g. oxygen is twice as viscous as hydrogen.) Liquids such as oil, syrup, honey, etc., have a high viscosity and when cold are more difficult to pour than when heated, when they become more runny. This is true of all liquids; as the temperature rises the viscosity decreases. Therefore the viscosity and the resistance in a warm water pool will be less than that in a cold water pool. The reverse is true of gases where the higher the temperature the greater the viscosity. The greater the viscosity of a liquid the greater the resistance offered to an object moving through it, as the greater the friction between the molecules of the liquid to the object and the surrounding molecules of that liquid. Viscous forces are very large at high speeds because the frictional forces are proportional to the velocity. The viscous forces are relatively small at slow speeds and therefore in hydrotherapy they will not be of great importance, especially if the water is warm and the viscosity is further decreased.

TURBULENCE. Objects travelling through fluids experience resistance to their motion. This varies according to the shape of the object and it arises from three different causes:

1   *Viscosity*. When an object travels through water a boundary layer of water is dragged along with it. Between this layer and the undisturbed water is friction which needs energy to maintain it.

2   *Waves*. These occur on the surface. Waves are generated by an object such as a ship as it passes through the water. These waves are of two kinds:

a   Waves tailing behind the object which travel forwards and move with the same velocity as that of the moving object,

b   waves formed at the front of the moving object producing the familiar V-shaped pattern. In deep water a ship moving at moderate speed will always produce V-shaped waves at the bow of 39°. Both these wave formations will need energy to maintain them.

3   *Turbulence*. For an object which is completely immersed in a fluid, this is the most important cause of resistance, and is the series of eddies following in the wake of an object. When an object moves slowly through water the water particles flow parallel or nearly parallel to the line of flow. (Figure 90.) Thus the water flows past the object in smooth continuous curves and this is called streamlined flow.

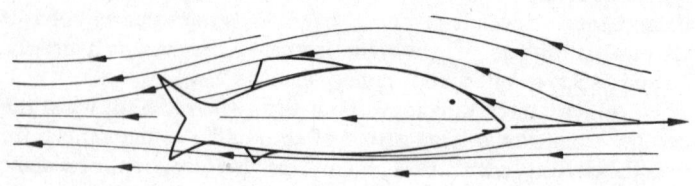

Fig. 90

Streamlined Flow is a continuous steady movement of a liquid, the rate of movement at any fixed point remaining constant. It can be pictured as very thin layers of fluid molecules sliding over one another and therefore friction between the layers is minimal. If a liquid flows through a pipe in streamlined flow the inner layers will move relatively faster than the outer layers, and the outermost layers will be almost stationary.

As the speed increases, eddies form behind the object and the energy which these eddies contain will be derived from the movement of the object and therefore the formation of eddies dissipates energy. This formation of eddies is called *Turbulent Flow*, and is produced when the velocity of the flow is increased beyond a certain level called the critical velocity. In turbulent flow there are rapid random movements of the molecules producing rotation in a clockwise and anti-clockwise direction. (Figure 91.) The rate of movement

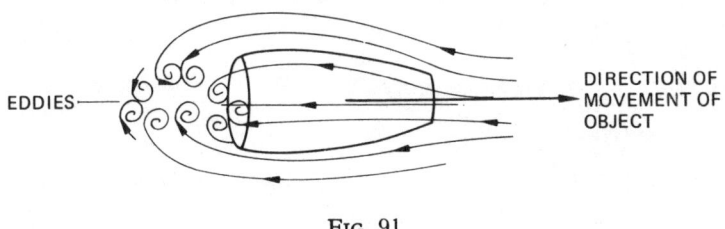

FIG. 91

at any fixed point will not be constant. Frictional resistance due to turbulent flow is greater than that due to streamlined flow. In streamlined flow the resistance is proportional to the velocity. In turbulent flow, the resistance is proportional to the square of the velocity. Streamlined flow is therefore the resistance offered between layers of fluid molecules and turbulent flow is the resistance due to friction between individual fluid molecules.

The production of eddies is greatly influenced by the shape of the body moving through the fluid. The less streamlined the more vigorous the eddies. A body rounded at the front and tapered behind, such as a submarine or a fish, travelling at the same speed as a bulky object with the same volume will produce far less frequent and vigorous eddies. The effect of turbulence is to produce a low pressure area immediately behind the moving object. This creates a suction effect which tends to pull the object back.

When a person is moving through water, turbulence can be reduced by streamlining the body so that he lies horizontal to the water surface with the arms close to the trunk and the legs, ankles and feet close together. Turbulence is used as a force of resistance in pool work. The quicker the movement the greater the turbulence, and therefore an exercise may be made more difficult by speeding up the movement. Turbulence can be further increased by making use of table-tennis-type bats which, when moved through the water flat side forwards, will produce a large number of eddies behind them.

DIFFUSION

If a strong salt solution is separated by a membrane permeable to salt and water, from a weak solution, salt passes from the strong to the weak solution, and so in time the solution on both sides of the

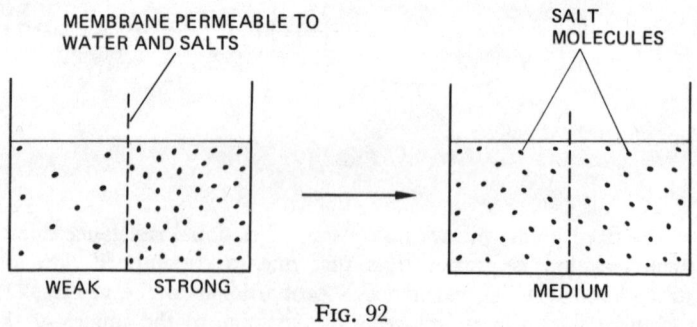

Fig. 92

membrane will have the same strength. This is known as diffusion (Figure 92). Diffusion is the random movement of molecules from an area of greater concentration to an area of lesser concentration.

OSMOSIS

Osmosis is the passage of water from a low to a high solute concentration through a semi-permeable membrane. If water containing a solute (i.e. a substance which is dissolved in it, such as sugar) is separated from pure water by a membrane permeable to water but not to the solution molecules (i.e. semi-permeable) the water molecules have an average activity higher on one side of the membrane than on the other. Molecules from the pure water cross the membrane with greater frequency than those from the solution. Since there is a quantitative difference between these movements, there is net movement of water molecules from the pure water to the solution. (Figure 93.) This passage of water through a semi-permeable membrane from a lower to a higher concentration is termed osmosis.

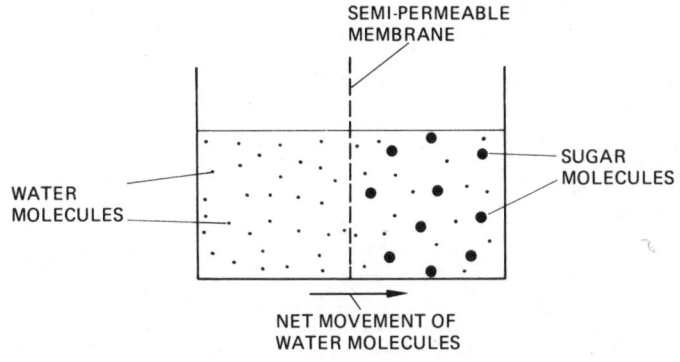

FIG. 93

Movement of water in this way will produce a difference in surface level and therefore in *hydrostatic pressure* between the two solutions. This pressure tends to oppose the entry of water and when of sufficient magnitude prevents further net movement across the membrane. The hydrostatic pressure required to prevent the net entry of a solvent across a semi-permeable membrane is called the Osmotic Pressure of that solution. (Figure 94). The osmotic pressure is proportional to the molecular or ionic concentration of the solution and therefore the more concentrated it is the greater the osmotic pressure it will exert. As a good example, prunes when dry have a high concentration of sugar in their cells, and are covered by a semi-permeable membrane. When soaked in water there is a net passage of water molecules into the prunes and they swell up. In the body, osmosis is a very necessary factor in the regulation of salts, water,

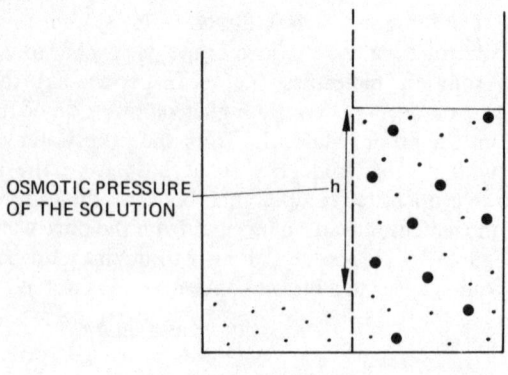

OSMOTIC PRESSURE
OF THE SOLUTION   h

Fig. 94

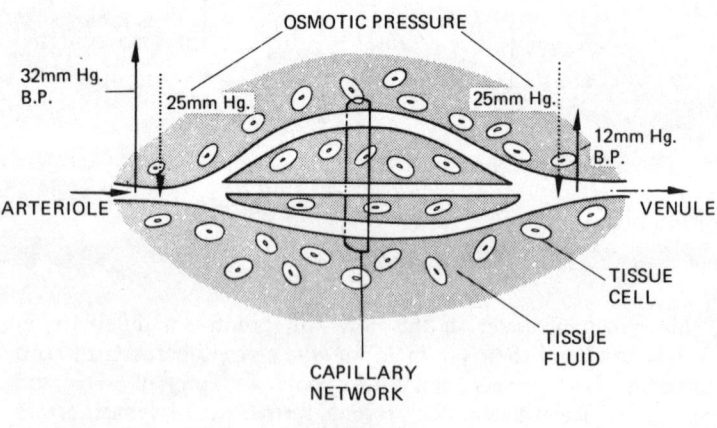

OSMOTIC PRESSURE

32mm Hg.
B.P.

25mm Hg.   25mm Hg.

                                                                          12mm Hg.
                                                                          B.P.

ARTERIOLE   VENULE

                                                                          TISSUE
                                                                          CELL

                                                                 TISSUE
                                                                 FLUID

                                CAPILLARY
                                NETWORK

Fig. 95

nitrogenous waste and carbon dioxide in the kidneys, lungs and skin. However, all living membranes are not truly semi-permeable because they allow the passage of ions and molecules, other than water, to pass through them. Selectively permeable would therefore probably be a better term.

The capillaries of the body are permeable to all the constituents of the plasma except the plasma proteins. Figure 95 is a diagrammatic representation of the capillary field. The blood pressure at the arterial end is 33mm.Hg. It has dropped to 12mm.Hg. at the venous end. The blood pressure will therefore tend to force fluid out through the capillary walls. However, due to the presence of proteins in the blood, but not in the tissue fluid surrounding the capillaries, an osmotic pressure of 25mm.Hg. is exerted which tends to suck back fluid into the capillary, so two opposing forces exist. At the arterial end the blood pressure is in excess of the osmotic pressure by 7mm.Hg., and therefore the fluid passes from the capillary into the tissues. At the venous end the osmotic pressure exceeds the blood pressure by 13mm.Hg., and therefore the fluid returns into the blood stream.

REFERENCES

ABBOT, A. F.: *Ordinary Level Physics.* (1968) Heinemann Educational Books Ltd., London.

ALEXANDER, R. M.: *Animal Mechanics.* (1968) Sidgwick and Jackson, London.

BRUNNSTROM, S.: *Clinical Kinesiology.* (1962) Blackwell Scientific Publications, Oxford.

CIBA FOUNDATION: *Myotatic, Kinesthetic and Vestibular Mechanisms.* J. A. Churchill, Ltd., London.

DUFFIELD, M. H. T.: *Exercises in Water.* (1969) Baillière, Tindall and Cassell Ltd., London.

DYSON, G.: *The Mechanics of Athletics.* (1968) University of London Press Ltd., London.

FREEMAN, I. M.: *Physics made Simple.* (1967) W. H. Allan, London.

GARDINER, M. D.: *The Principles of Exercise Therapy.* (1969) G. Bell and Sons Ltd., London.

GREEN, J. H.: *An Introduction to Human Physiology.* (1968) Oxford University Press, London.

GREENWOOD, M. E.: *An Illustrated Approach to Medical Physics.* (1966) F. A. Davis Company, Philadelphia.

GUYTON, A. C.: *Function of the Human Body.* (1969) W. B. Saunders Company, Philadelphia, London, Toronto.

HACKER, G. H.: *First Steps in Physics.* (1966) University of London Press, Ltd., London.

HOLLIS AND ROPER: *Suspension Therapy in Rehabilitation.* (1958) Baillière, Tindall and Cox Ltd., London.

JARDINE, J.: *Physics is Fun.* (Volume 3) (1969) Heinemann Educational Books Ltd., London.

LOCKWOOD, A. P. M.: *Animal Body Fluids.* (1966) Heinemann Educational Books Ltd., London.

NIGHTINGALE, A.: *Physics and Electronics in Physical Medicine.* (1959) G. Bell and Sons Ltd., London.

STEINDLER, A.: *Kinesiology of the Human Body.* (1955) Charles C. Thomas, Illinois.

TRICKER, R. A. R.: *The Science of Movement.* (1967) Mills and Boon Ltd., London.

WELLS, K. F.: *Kinesiology.* (1966) W. B. Saunders Company, Philadelphia and London.

The Student's Book of Understanding Science. (1963) Sampson Low, London.